Contract Farming and the Development of Smallholder Agricultural Businesses

Contract farming has received renewed attention recently as developing economies try to grapple with how to transform the agricultural sector and its associated value chains. This book examines different contract arrangements for selected crops, applying both qualitative and quantitative approaches in order to examine how contract farming affects smallholders and value chain dynamics in Tanzania.

Major themes covered in the book include: contract farming policy; contract farming and value chain dynamics; contract farming adoption decisions; contract farming and income diversification. The authors also discuss alternative aspects of contract farming such as trust, conspiracy, empowerment and corporate social responsibility. The book presents original research from case studies conducted in Tanzania on sugarcane, tobacco, sunflower and cotton. These crops have a history of trials and errors with contract farming involving smallholders. Futhermore, they are targeted in national strategies as some of the main crops for establishment and upgrading of agro-industrial activities in Tanzania.

Joseph A. Kuzilwa is a Professor in the Department of Economics, Mzumbe University, Tanzania.

Niels Fold is a Professor in the Department of Geosciences and Natural Resource Management, University of Copenhagen, Denmark.

Arne Henningsen is an Associate Professor in the Department of Food and Resource Economics, University of Copenhagen, Denmark.

Marianne Nylandsted Larsen is an Associate Professor in the Department of Geosciences and Natural Resource Management, University of Copenhagen, Denmark.

Other books in the Earthscan Food and Agriculture Series

Environmental Justice and Farm Labor
Rebecca E. Berkey

Plantation Crops, Plunder and Power
Evolution and Exploitation
James F. Hancock

Food Security, Agricultural Policies and Economic Growth
Long-term Dynamics in the Past, Present and Future
Niek Koning

Agriculture and Rural Development in a Globalizing World
Challenges and Opportunities
Edited by Prabhu Pingali and Gershon Feder

Food, Agriculture and Social Change
The Vitality of Everyday Food in Latin America
Edited by Stephen Sherwood, Alberto Arce and Myriam Paredes

Contract Farming and the Development of Smallholder Agricultural Businesses
Improving Markets and Value Chains in Tanzania
Edited by Joseph A. Kuzilwa, Niels Fold, Arne Henningsen and Marianne Nylandsted Larsen

Agribusiness and the Neoliberal Food System in Brazil
Frontiers and Fissures of Agro-neoliberalism
Antonio Ioris

For further details please visit the series page on the Routledge website: www.routledge.com/books/series/ECEFA/

Contract Farming and the Development of Smallholder Agricultural Businesses

Improving Markets and Value Chains in Tanzania

Edited by
Joseph A. Kuzilwa, Niels Fold, Arne Henningsen and Marianne Nylandsted Larsen

Routledge
Taylor & Francis Group
LONDON AND NEW YORK

earthscan
from Routledge

First published 2017
by Routledge
4 Park Square, Milton Park, Abingdon, Oxon OX14 4RN
605 Third Avenue, New York, NY 10017

Routledge is an imprint of the Taylor & Francis Group, an informa business

First issued in paperback 2019

British Library Cataloguing in Publication Data
A catalogue record for this book is available from the British Library

Library of Congress Cataloging in Publication Data
A catalog record for this book has been requested

ISBN: 978-1-138-12074-7 (hbk)
ISBN: 978-0-367-35134-2 (pbk)

Typeset in Bembo
by Wearset Ltd, Boldon, Tyne and Wear

Contents

Figures

Tables

Contributors

Frederik Brønd, PhD fellow, Department of Geosciences and Natural Resource Management, University of Copenhagen, Denmark.

Niels Fold, Professor, Department of Geosciences and Natural Resource Management, University of Copenhagen, Denmark.

Jonas Gillett, MSc in Geography and Geoinformatics, Ministry for Environment and Food, Agency of Water and Nature Management, Denmark.

Arne Henningsen, Associate Professor, Department of Food and Resource Economics, University of Copenhagen, Denmark.

Bahati M. Ilembo, Post-doctoral fellow, Department of Economics, Mzumbe University, Tanzania.

Lotte Isager, Post-doctoral fellow, Department of Geosciences and Natural Resource Management, University of Copenhagen, Denmark.

Joseph Andrew Kuzilwa, Professor, Department of Economics, Mzumbe University, Tanzania.

Marianne Nylandsted Larsen, Associate Professor, Department of Geosciences and Natural Resource Management, University of Copenhagen, Denmark.

Daniel Mpeta, Lecturer, Institute of Rural Development Planning, Dodoma, Tanzania.

Paul Maganga Nsimbila, Lecturer, Department of Procurement and Logistics Management, School of Business, Mzumbe University, Tanzania.

Thobias E. Nsindagi, Lecturer, Department of Economics, Mzumbe University, Tanzania.

Batimo Sebyiga, Senior Lecturer, Institute of Rural Development Planning, Dodoma, Tanzania.

Jennifer K. Sesabo, Lecturer, Department of Economics, Mzumbe University, Tanzania.

Acknowledgements

This book could not have been completed without the hospitality and responsiveness from the many farmers scattered in the rural areas of Tanzania where we carried out the study. They patiently listened and answered our questions about their agricultural practices, perception of contract farming and livelihood strategies. This information constitutes the most crucial part of the book and we are grateful for their willingness to help us with the collection of data on the contract farming schemes we have studied.

The research has benefited from generous financial support from Danida, Denmark's development cooperation under the Danish Ministry of Foreign Affairs (Grant no. 10-P11-TAN). Funding has even continued after the completion of the first research project so a second phase of the research on contract farming in Tanzania is currently going on – with many of the same researchers involved. We are also grateful to the Danida Fellowship Centre who provided much appreciated administrative support and kindly hosted the Tanzanian PhD scholars during their long-term study stays in Denmark. We also wish to thank the Tanzania Commission for Science and Technology (COSTECH) who expediently issued the necessary research permits to the Danish researchers.

Needless to say, the writing of this book has received the help and assistance of countless of other people during the project period 2011–15 and we can mention only a few. In the field, Lington Kiaro and Cathbert Mwanyika provided invaluable research assistance and uncomplainingly acted as Swahili translators for the Danish team members. We also gratefully appreciate the support of all enumerators from Mzumbe University and Institute of Rural Development Planning (Dodoma), who diligently assisted in data collection during the surveys. In Denmark, our thanks go to Kent Poerksen for the map drawings, Robert Parkin for efficient and careful language revision, and Sinne Broby Oertenblad for her thorough work with the proofreading and formatting of the chapters before submission to the publisher. We also wish to acknowledge and thank Henrik Hansen and Jens-Martin Bramsen for their dedicated participation in the early phases of the project.

Being so heavily based on rich empirical material, the book has benefited enormously from the many informants who kindly shared their expertise with us without expecting anything in return. Obviously, each of them is deeply implanted in one of the crops that were studied – apart from the assistance we got from the Ministry of Agriculture, Livestock and Fisheries. For *sugar* we wish to thank representatives from Tanzania Sugar Board, Mtibwa Sugar Estate, the out-growers' associations and the cooperative societies who helped us link up with the smallholder sugarcane-growers. For *cotton*, we are indebted to representatives from a number of private ginneries, Tanzania Cotton Board, Tanzania Cotton Grower Association (TATOGA), Tanzania Cotton Association, the Cotton Development Trust Fund, TechnoServe, Tanzania Gatsby Trust and not least many of the leaders of farmer business groups. For *tobacco*, we acknowledge the cooperation by the District Executive Directors (Offices of District Agricultural, Irrigation and Cooperatives) for logistical support and for linking us with the grass-roots-level officials in the tobacco growing districts under study. Many thanks are also due to the officials of the primary cooperative societies, the cooperative unions, the Association of Tanzania Tobacco Traders, tobacco processing factories and the Tanzania Tobacco Board. For *sunflower seed*, we wish to thank the owners of small and medium-sized sunflower oil processing companies for allowing us to visit their facilities and the invaluable information and practical support during the fieldwork. Moreover, we learned a lot and received practical support and advice from representatives of the Rural Livelihood Development Company (RLDC), the Sunflower Processors Association (TASUPA), the Tanzania Edible Oil Seeds Actors Association (TEOSA) and the Central Zone Sunflower Oil Processing Association (CEZOSOPA).

We are thankful for comments and feedback on earlier drafts of the chapters from participants of the POLICOFA International Conference on Contract Farming in Africa (Dar es Salaam, October 2014). While we take full responsibility for the ultimate result, their comments were of considerable help in our endeavours to improve the chapters. In the final phases we have been fortunate to have very patient and understanding support from Routledge's Tim Hardwick and Amy Johnston who carefully saw the book to the end.

Finally, we would like to sincerely thank our colleagues in the project team who remained truly dedicated to the book project and, with a good sense of shared trans-continental humour, accepted our repeated requests for revisions. It has been an enlightening and enjoyable experience to carry out the editorial work among Tanzanian and Danish partners in a highly collaborative spirit.

Part I

Contract farming in context

Part I

1 Researching the potential and limitations of contract farming in sub-Saharan Africa

Joseph Andrew Kuzilwa, Niels Fold, Arne Henningsen and Marianne Nylandsted Larsen

Introduction

Since political independence, most countries in sub-Saharan Africa have struggled to improve the welfare of their people by increasing production and enhancing market access for agricultural products. The rationale behind these efforts is straightforward: more than 60 per cent of the population in sub-Saharan Africa live in rural areas, the majority being smallholders with agriculture as the main component of their livelihoods, even though non-farm incomes are becoming increasingly important (Bryceson, 1996). Growth in agricultural production has a significant effect on economic growth: A cross-country study in Africa suggests that agriculture is the main engine of growth and the sector with the highest potential for poverty reduction (Diao *et al.*, 2007).

Historically, different ways of organising production and market arrangements have been attempted in Africa, including direct involvement by the state in agricultural marketing. Following the economic liberalisation and structural adjustment reforms of the 1980s and 1990s – and in particular after the gradual dismantling of state-controlled marketing systems (parastatals) – the private sector has moved to play a more direct role in agricultural production and marketing. The economic liberalisation has resulted in macroeconomic growth with new opportunities for the marketing of consumer goods (including agricultural goods) within national economies, as well as on the world market, as new export-oriented strategies were implemented as part of structural adjustment programmes. The new opportunities have resulted in a much broader incorporation of multiple stakeholders, who cooperate and compete within the confines of global and national agribusiness value chains. From a North–South perspective, these stakeholders are dominated by huge transnational companies that operate in the processing and retailing segments of the agricultural value chains, but they also include smallholders and local firms seeking to meet the national and global demand for high-value agricultural products.

Although various forms of contract farming arrangements have existed parallel to other market arrangements since the initial period of independence, contract farming as a strategic and developmental element has received far greater attention in sub-Saharan Africa during recent decades (for Tanzania, see TNBC, 2009; URT, 2010, 2016). Contract farming is now considered a policy approach capable of addressing commercial demand from national and international consumers while also addressing the problems of low productivity, limited market access and inadequate farm incomes among smallholder farmers in developing economies. Hence, by integrating systems of agricultural production, extension and marketing, in principle contract farming enables smallholders to access international product markets (Birthal *et al.*, 2007; Barrett *et al.*, 2012). Products traditionally not considered suitable for production under contract farming are increasingly being incorporated into this type of arrangement, one example being sunflower seeds produced by smallholders in Tanzania.

The increased use of contract farming arrangements for smallholders in an expanding number of (non-traditional) crops has gone hand in hand with a renewed research interest in the extent to which such arrangements deliver all or some of their intended outcomes, like improved productivity, higher income and overall welfare, as well as how they impact on the poorer farmers who are generally the weak partners in such contractual arrangements (Barrett *et al.*, 2012). Research findings have so far not been unequivocally conclusive about the assumed benefits of contract farming, especially regarding welfare and the range of different impacts on smallholders (Bellemare, 2015; Oya, 2012; see also Chapter 2, this volume).

The present collection examines the potentials and challenges of contract farming in Tanzania. It is based on research undertaken in different sites but focuses on four crops produced under different forms of contract farming arrangements. The research was carried out between 2011 and 2015 within the framework of a research programme financed by the Danish International Development Agency (Danida) under the Ministry of Foreign Affairs (Grant no. 10-P11-TAN). The overall objective of the book is to present research findings on the role and importance of contracts for the small-scale farming of sugarcane, tobacco, sunflower and cotton in Tanzania. The selected crops have different characteristics and histories. Two of them, tobacco and sugarcane, have a relatively long history of 'trial and error' with contract farming involving smallholders, while the other two crops, sunflower and cotton, are comparatively new to contract farming. Moreover, cotton and tobacco are so-called industrial crops, while sugarcane and sunflowers are food crops, although both must be processed for human consumption as sugar and edible oil respectively. The purpose of including these different crops in the research programme was to examine the different ways in which the 'nature' of the crop conditions the organisational structures and dynamics in the concomitant value chains. The aim was to take as our point of departure a common theoretical platform and methodology in order to offer a wider perspective on contract farming that includes several different crops and sites in Tanzania: sunflower in Kongwa District, sugarcane in Mvomero and Kilombero Districts, cotton in Bunda District and tobacco in Urambo District (see Figure 1.1).

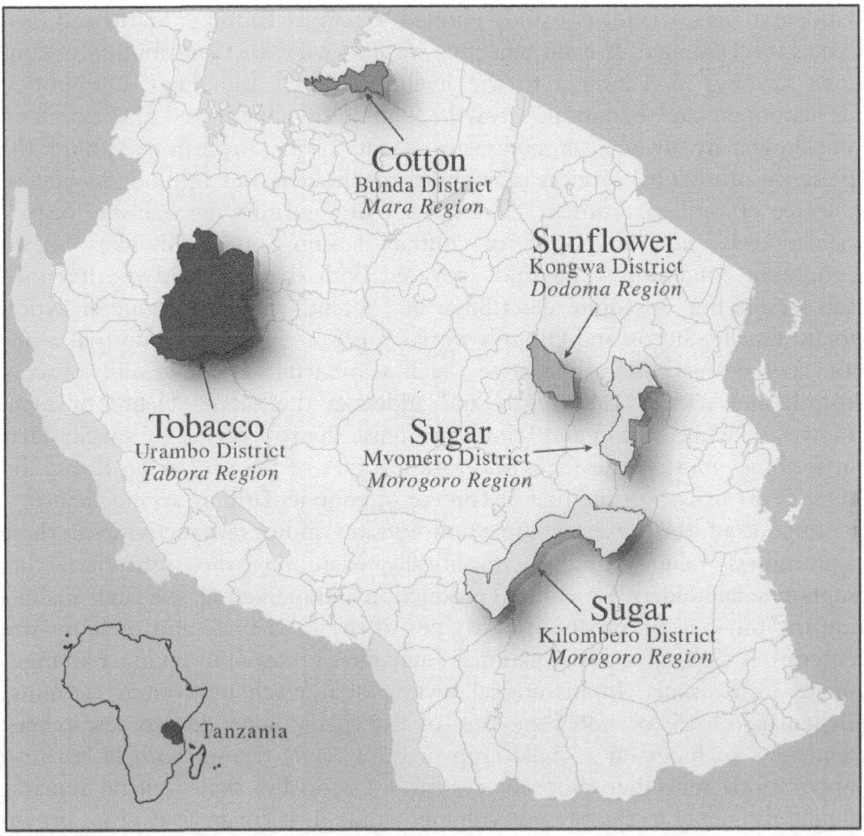

Figure 1.1 Crops and study districts in Tanzania.

Research on related issues in Tanzania is quite limited, both in terms of the number of crops covered and the level of detail. Three examples serve to support the rationale of this book. One of the first studies of contract farming was the seminal work by Boesen and Mohele (1979), who used a historical approach to trace the emergence of tobacco contract farming in Urambo since the early 1950s. Tobacco was introduced after the failure of a groundnut scheme in the area. It was initially started by European settlers but was subsequently successfully adopted by local smallholders in the early 1960s to such an extent that tobacco became one of the major export crops and an important source of livelihood for thousands of households in Tabora. While the authors provide a history of tobacco production in Tanzania and of the licensing of farmers to grow the crop, it was written too early to have covered the recent development of contract farming arrangements in this sector. In 2006, under the World Bank's Participatory Agricultural Development and

Empowerment Project (PADEP), the Ministry of Agriculture, Food and Cooperatives carried out a study entitled 'Contract Farming: Status and Prospects for Tanzania'. A main objective was to review the contribution of contract farming in Tanzania to the livelihoods of smallholders, agribusiness development and economic growth. The study illuminated the extent of smallholder involvement in contract negotiations (e.g. over price setting), the presence of essential services to support smallholders and agribusiness buyers, the role of local government, the existence of regulatory mechanisms for particular crops and the impact of contract farming on smallholders' socio-economic conditions. The study covered both the crop and the livestock sub-sectors but was quite descriptive in content, with an opaque analytical focus. Finally, Sutton and Ulomi's work (2012) on Tanzania's industrial sector also includes various agribusinesses, such as oilseeds and edible oils, tobacco, cotton and textiles, and sugar, and addresses the supply chains of each. However, while Sutton and Ulomi recognise the role played by smallholders in feeding industry, the focus is on the profiles of the processing firms, and there is no analysis of the role or content of contract farming arrangements.

In contrast, the present volume concentrates on upstream activities in these agribusiness value chains by specifically examining the institutions that support smallholders' agricultural production and marketing, the latter including the buying and processing activities of the local small and medium-size enterprises (SMEs) and transnational companies involved in contract arrangements in Tanzania. In his critical review of research on contract farming, Bellemare (2015: 3) notes the lack of external validity due to researchers' tendency 'to focus on a single crop or on a single region, with little to no implications for other crops or regions'. Essentially, these critical remarks might also apply to the present volume: although it covers four crops grown in five different locations in Tanzania, the overall regulatory framework is identical in all studies. On the other hand, the coverage of four crops with different crop characteristics and histories demonstrates the heterogeneity of contract farming arrangements (Barrett *et al.*, 2012). It is therefore hoped that the book will contribute to current issues, especially the contract farming strategies adopted by emerging African economies. In that sense, the focus on a single country is an advantage, as it makes it possible to assess the implications of contract farming and align them with specific national strategies and plans within a theoretical framework grounded in economic contract theory and global value chain analysis. In the following sections, the common theoretical framework is elaborated and the contents of the different empirical analyses are outlined.

Theoretical perspectives I: economic contract theories

A systematic study of contract farming needs to be rooted in a well-grounded theoretical framework. Contract farming is related to many economic theories. For instance, farm production under contract can be analysed using tools from microeconomic production theory (see e.g. Chambers, 1988). As contract farming is often carried out by small-scale farmers, with many resources being provided by the farmers' households (e.g. land, labour, liquidity), and as several input and output markets are affected by market imperfections, (contracted) farm production is closely interlinked with household decision-making. Under these commonly observed circumstances, microeconomic household theory and (agricultural) household models (see e.g. Singh *et al.*, 1986) are required to analyse certain aspects of contract farming. Contract farming arrangements often include elements of risk-sharing between the farmer and the contractor so that analysis of these contracts needs to take into account theories of decision-making under uncertainty (see e.g. Chambers and Quiggin, 2000). Transaction cost theory is another economic theory that is important for analysing certain aspects of contract farming, because contract farming aims at reducing transaction costs by setting up a 'hybrid' governance structure occupying the middle ground between a 'market' and a 'hierarchy' (vertical integration) (see e.g. Williamson, 1991). As contract farming entails contracts, which establish interdependencies between the contractors and the contracted farmers, contract theory, and particularly its sub-field of principal–agent theory (see e.g. Bolton and Dewatripont, 2005; Laffont and Martimort, 2002), are necessary for analysing several important aspects of contract farming. Finally, as contract farming may affect the welfare of different groups of people, either positively or negatively, one can analyse the implications of contract farming using tools taken from welfare economics.

As several chapters in this book are based on principal–agent theory (e.g. Chapters 8 and 9), we briefly elaborate on some aspects of this theory that are relevant for contract farming arrangements. The basic principal–agent model depicts the problems or difficulties of establishing contractual relationships. It takes into account the existence of informational problems in the contracting process between two self-interested parties with divergent interests. A principal–agent relationship exists, for instance, if there is a contract under which a contractor ('the principal') engages a contract farmer ('the agent') to produce a crop on his or her behalf, thus involving the delegation of some decision-making authority from the principal to the agent (Gardner, 2003). This delegation, however, also means that the principal must place trust in the agent to act in his or her interests. For instance, if a contractor provides inputs to a contracted farmer, there must be trust that the contract farmer will use these inputs for the contracted crop and will sell the harvested output to the contractor. If the contract farmer is a (short-run) utility maximiser, he or she may not act in the best interests of the contractor, but use some of the inputs for other crops ('input diversion') and/or sell the harvested output to a different buyer ('side-selling').

These problems can occur because principal–agent relations such as contract farming arrangements are usually characterised by information asymmetries: that is, at least some information is known to one party, but not to all parties involved in the contractual process. Self-interested agents can use this information asymmetry to their own advantage and to the disadvantage of the principal. Therefore, the principals usually lack reasons to trust the agents and will make efforts to resolve this concern by designing incentive-based contracts that can induce the agent to undertake actions that will maximise the principal's welfare even in the latter's physical absence. Principal–agent theory can be used to design optimal incentive schemes so that contracts align the interests of the agent with those of the principal, thus reducing the scope for information asymmetries and opportunistic behaviour (Shapiro, 2005).

The problems that can occur in contract farming arrangements can often be attributed to 'adverse selection' or 'moral hazard'. Adverse selection occurs if opportunistic self-interested farmers who intend from the outset not to comply with a contract for their own benefit have much greater incentives to join contract farming than honest farmers who intend to comply. In this case, the proportion of opportunistic farmers can be much larger in the group of contracted farmers than in the population of all farmers so that a large proportion of contract farmers are not complying with the contract but practising, for example, input diversion or side-selling. In contrast, moral hazard occurs if farmers initially intend to comply with the contract but change their intentions and behaviour after the contract has been signed and becomes operational, thus ultimately failing to comply with the contract.

A crucial assumption of the general principal–agent model is that individuals are depicted as being rational, that is, they are assumed to be self-interested, to exploit the situation they face for their own benefit and to choose the best means to pursue their goals given the information they have. Thus, an individual will observe a contractual commitment in the contract only to the extent that it is of benefit, otherwise he or she will seek to improve his or her position by withholding information, misreporting or not acting according to the agreement (e.g. Williamson, 1996). However, this assumption is not fulfilled in many empirical cases, as it has been shown that people often do not behave as if they were only interested in their own short-term advantages (e.g. Ostrom, 1998).

Another assumption of the general principal–agent model is that all future contingencies can be foreseen and that contracts provide solutions to all contractual problems, including all future contingencies. There is no need to renegotiate contracts, and there is no need for interventions of third parties like the government, as these would constrain some activities and lead to inefficiencies (Wu and Roe, 2007). However, as individuals usually cannot foresee all possible future contingencies, they cannot specify the actions to deal with these contingencies in their contracts. And even if it were possible, the costs of making and administering such contracts would be excessive. Therefore, contracts are at any one time incomplete, and re-negotiation will often be required (Hart, 1995, as cited in Bogetoft and Olesen, 2004; Wu and Roe, 2007).

Furthermore, principal–agent theory does not address some important aspects of contract farming arrangements, such as coordination, motivation and cost minimisation, and it also fails to provide sufficient guidance for understanding the multiple decisions that need to be made in practical contracts (Bogetoft and Olesen, 2004). Bogetoft and Olesen (2004) propose a holistic rather than a partial approach for understanding contractual relationships in agricultural production. In this approach, a contract is based on the conceptual framework of multi-criteria decision-making, which requires various aspects of the contract to be taken into account simultaneously (Bogetoft and Olesen, 2004).

In either a holistic approach to contract farming (Bogetoft and Olesen, 2004) or a complete approach to contract farming (Wu, 2006), it is argued that there may be other reasons than pure economic self-interest why parties sign a contract: for example, to maximise the integrated gain (i.e. sum of profits of all contracting parties). The intuition behind seeking the maximum overall gain is that a large overall gain may make each party better off, provided it is possible to redistribute the overall gain or profit without undesirably affecting the favourable behaviour of the contracting parties. Contracts leading to such maximum unified profits are called Pareto-efficient contracts or complete contracts (Wu, 2006). These are desirable contracts that can be achieved through bargaining if property rights are well defined. According to Bogetoft and Olesen (2004), there are three fundamental roles that these contracts must fulfil: coordinate production and marketing, motivate the parties and minimise transaction costs.

Theoretical perspectives II: global value chain analysis

The inter-connection of people and places has increased enormously with the expansion and intensification of globalisation in recent decades. Places are progressively becoming interlocked through flows of goods, finance, labour and information, if not directly through straight linkages, then indirectly through intricate and complex webs of commercial and social relations. Hitherto mutually isolated places are now tied together in interlinked development trajectories. For some geographers, development is essentially about these interlocking relationships between places and people – how they are established, consolidated and possibly eroded, and how they impact on the 'lived experience' of poor and marginalised people through multi-scalar interactions (Lawson, 2007).

A useful entry point to improve understanding of the dynamics behind these processes of inter-connectedness is chain-inspired analyses that link producers of raw materials with consumers of final goods. This approach is not an analytical panacea for all research themes under the 'globalisation umbrella' mentioned above. However, maintaining a product-specific focus while mapping nodes (i.e. points of transformation and value-adding) and actors involved in the chain is a tangible way to start examining particular themes and issues in parts (or indeed the whole) of a chain (Fold, 2014). We argue that contract farming for smallholders has to be positioned within a wider understanding of the structures and dynamics embedded in the particular agro-industry in which they unfold. Contract farming schemes are affected by broader regulatory mechanisms (both nationally and internationally) and economic dynamics on the world market, and the contractual relationships between smallholders and buyers or processors do not take place in isolation from these processes.

The chain-related analytical tradition has its roots in World Systems Theory, which was preoccupied with the causes of historically and cyclically determined transformations of commodity chain structures (Hopkins and Wallerstein, 1994). In this line of thought, different production processes are organised within distinct segments (or 'boxes' in the original terminology), but connected through input–output relations between the segments. Different chain segments are attached to countries and places, which in turn are predominantly located at the core, periphery or semi-periphery during certain periods. This approach was reformulated into the Global Commodity Chain (GCC) framework by the sociologist Gary Gereffi and associates (Bair, 2005, 2009). Avoiding the rigid geographical and conceptual distinction between core, periphery and semi-periphery, the GCC approach was developed in an attempt to understand and analyse the dynamics of the present phase of global capitalism, with its increased and diversified worldwide flows of capital and goods (Gereffi, 1994, 1995). A GCC is constituted by four elements: (1) an input–output structure or configuration, that is, a set of economic activities linked together in a value-adding sequence; (2) a specific geography reflecting the spatial distribution of networked enterprises; (3) an internal governance structure that determines how resources are allocated and flow within the chain; and (4) a set of public and private institutions that operate at different spatial scales and regulate the actors and their interaction (see Neilson and Pritchard, 2009, for a discussion of the interrelationship between governance and institutional framework).

Chain governance – the allocation of resources and non-market coordination of activities within the chain – was singled out as a key theoretical concept. Governance is considered necessary in order to meet market demand (related to product design) and process specifications, as well as to reduce the risks of not complying with strict adherence to requirements concerning quality, reliability and response times for supplies (Humphrey and Schmitz, 2002). Governance involves the ability of a so-called lead firm to determine the activities of other companies in the chain. Lead firms have the power to define products, specify procedures and standards, and control vital resources needed in the chains. Hence, chain governance can be identified by examining the practices of the lead firms that are the dominant actors in the chain. Governance, it was argued, was revealed in two prevalent types of global commodity chains, producer-driven and buyer-driven chains. In buyer-driven chains, retailers and brand name companies that control R&D, design and marketing set the parameters that are necessary for coordination of the chain, while in producer-driven chains the key parameters are set by the companies that control the main product and processing technologies.

Originally, the producer- and buyer-driven chains were developed as ideal types and superimposed on (1) two types of industrialisation strategy based on import substitution and export orientation (ISI vs. EOI), and (2) two types of industrial organisation (mass production vs flexible specialisation) (Fold and Larsen, 2008). Hence, the theoretical foundation of these two forms of governance is rather rudimentary, and their 'logics' are firmly linked to chains in which industrial processing is crucial.

After the turn of the present century, the concept 'officially' changed from GCC to Global Value Chain (GVC): 'commodity' was replaced with 'value' to better capture the variety of products (but still including those with the nature of commodities) and signify the importance of value-adding processes (Gereffi *et al.*, 2001). Moreover, critics argue that the ideal types of chain governance, dichotomised between buyer- and producer-driven, were empirical generalisations rather than theoretical constructs (Humphrey and Schmitz, 2002). Instead, Gereffi *et al.* (2005) proposed a set of governance forms on the basis of the relationship between lead firms and first-tier suppliers. The authors argued that, in addition to market and hierarchy, lead firms organise their first-tier suppliers in three different intervening forms of networks, through which they are able to deal with asset specificity, opportunism and coordination costs in repeat transactions. The key parameters for determining the actual form of governance are the characteristics of the transaction (complexity, codifiability) and the capability of the supplier, parameters that are further assessed as high or low.

Thus the high complexity and codifiability of transactions, combined with low supplier capability, results in captive governance, as lead firms offer potential suppliers a package of technical, financial and managerial assistance to enable them to take on board various low-profit service functions linked to the actual input production. Suppliers are effectively controlled but not owned by lead firms (Dicken, 2015). In contrast, relational governance is characterised by high capabilities among suppliers but low codifiability of highly complex transactions. The relationship is more egalitarian, and power asymmetries are not distinct, as with captive governance. Suppliers may have acquired knowledge and developed proprietary technology putting them in a better position to negotiate commercial conditions with buyers, or they may belong to ethnically based 'transnational communities'. Often the spatial proximity of the actors involved is essential in maintaining this type of governance. Finally, supplier capability and the complexity of transactions are high in modular governance, but here codifiability is also high. Hence, the complex information necessary for the implementation of the transaction can be 'condensed' into concise standards and other types of codified information, although this requires advanced knowledge and technological capacity among suppliers to carry out actual production. These types of suppliers, often denoted contract manufacturers or turnkey suppliers, have developed expertise in mastering flexible base processes or base components that can be adapted according to different buyer specifications while also exploiting economies of scale.

Analysing governance is crucial to understanding how lead firms organise their first-tier suppliers and thereby affect the dynamics of GVCs. In principle, however, the examination of buyer–seller relationships can be carried out along the complete GVC, that is, between actors in separate but linked boxes or segments. As a consequence, some observers distinguish between different forms of coordination between boxes on the one hand and the overall chain 'drivenness' or governance of the GVC on the other hand (Gibbon and Ponte, 2005). In addition, concerns were raised whether different forms of coordination and governance are prevalent in each of the filaments that constitute the chain, that is, whether distinguishing governance in possible sub-chains is necessary due to distinctive end-markets, actors and commodity characteristics (Gibbon, 2008). An example is whether robusta and arabica coffee constitute different filaments or are interlinked to such an extent that they are part of the same GVC.

A significant developmental dimension was added to GVC analysis with the introduction and adaptation of the concept of upgrading. In the early days, this concept was defined and used rather confusingly as an amalgam of improvements of the GVC position for firms, industries, countries and regions (see Gereffi, 1999), that is, 'getting a better deal' for the particular economic or territorial unit in question by participating in GVCs. A more concise definition and suggestion for analytical adaptation was based on a fourfold categorisation linked to the level of the firm (Humphrey and Schmitz 2002). Following this classification, process upgrading takes place when a product is produced more efficiently, product upgrading when production results in a product with a higher unit value, functional upgrading when production expands in scope and activities, and chain upgrading when acquired competences in one GVC are transferred to another GVC, resulting in an improved position. Upgrading and governance are intrinsically related, as some forms of governance promote or reduce upgrading potentials (Dolan and Humphrey, 2004).

Although these four categories are seemingly clear and easy-to-use concepts, empirical studies have revealed that they are not straightforward to apply. A lack of analytical clarity to reflect real-world phenomena has been observed (Ponte and Ewert, 2009; Selwyn, 2013), and theoretically as concepts they are not well developed. This is because they put too much emphasis on the buyer's willingness and capacity to upgrade the supplier, they are not open to combined and 'autonomous' upgrading efforts by the supplier and they implicitly establish a hierarchical order starting from process to chain upgrading (see above), even though substantial rewards (in terms of absolute profit at the firm level) are sometimes obtained by 'downgrading', i.e. moving to a specialised position in a 'lower'-ranked upgrading category (Gibbon, 2008; Tokatli, 2013). Nevertheless, recent studies have tried to incorporate the critical remarks and have thus improved the original concepts. The upgrading categories have now been adapted to empirical realities to take account of simultaneous and autonomous upgrading strategies resulting in multiple upgrading trajectories at the firm level (Hansen *et al.*, 2016).

For some observers, this 'technical' discussion of upgrading at the firm level misses the much broader issue of upgrading for whom? This issue has both a social and a territorial dimension (Bair, 2009; Barrientos *et al.*, 2011; Tokatli, 2013). First, it is questionable whether successful upgrading at the firm level has any spillover effects on the people employed, notably the workers in agriculture and manufacturing industries. Obviously, the sheer creation of new jobs in an earlier phase has the potential to socially upgrade employees who may have earned a lower or no income before being hired. But there are no automatic mechanisms to ensure that profits from upgrading are distributed to all employees so that the gains spread to the wider economy through increased purchasing power and increased demand for basic consumer goods. Moreover, although some of the permanently employed workers may benefit in terms of better wages and more secure working conditions, benefits in the form of both measurable standards (wages, working time, etc.) and enabling rights (freedom of organisation, right to strike, etc.) rarely spread to the increasing mass of cheap, temporary and 'flexible' labour employed in agriculture and industry (Barrientos *et al.*, 2011). Second, territorial progress in the form of regional development is not one of the upgrading categories: while some areas may benefit from participation and upgrading in GVCs, others areas may be marginalised and experience economic stagnation or even decline (Coe and Hess, 2011). Hence, the concept of upgrading has many connotations, of which the economic and firm-focused aspect is only one, albeit important, connotation, which is intimately related to both social and territorial upgrading.

Finally, some observers have remarked that the GVC and upgrading concepts are devoid of contextual and culturally conditioned intra-chain linkages. In a rival approach, known as the Global Production Network (GPN), these linkages are conceptualised as territorial and network embeddedness (Coe *et al.*, 2008; Coe and Hess, 2011). Territorial embeddedness encompasses the ways in which GVCs influence and are influenced by specific places (particularly through the presence of lead firms), while network embeddedness denotes how actors in a specific GVC are related to each other. The latter in particular offers a way of incorporating culture into chain-inspired analysis. Tight and sustained networks are based not only on trust and coalitions, but also on the existence of mutually congruent interests distributed in asymmetrical power relations, like, for instance, those between smallholders and local SMEs or transnational agribusiness firms.

Overview of the book

Taking their starting points in these two theoretical platforms, the following chapters address a number of 'traditional' issues pertaining to market access and income effects in relation to contract farming, but they also deal with contract farming from within the wider perspective of value chain coordination and dynamics. Moreover, a number of 'non-traditional' issues are analysed within the framework of contract farming: namely, issues like trust, conspiracy, empowerment and corporate social responsibility. The book is divided into four parts.

Having presented the aims and key theoretical entry points of the book in this chapter, Part I continues with a review of previous conceptualisations of contract farming and analytical foci. In Chapter 2, it is argued that recent contributions are dominated by econometric studies of contract farming's impact on household income, while broader issues of power and stratification within and among households have only been studied to a limited extent. Instead, renewed attention needs to address the consequences of contract farming for local or regional socio-economic dynamics. Finally, the authors point out the peculiar relegation of the contract itself to the sidelines of scholarly research.

Chapter 3 presents an overview of the agricultural sector in Tanzania, focusing particularly on the historical development of agricultural marketing systems since independence, and it briefly sets the scene for the four crops (sugarcane, tobacco, cotton and sunflower seed) that are examined in this book. The chapter traces the emergence and consolidation of contract farming in the government's various official policy documents and shows how the concept is increasingly being stressed as a tool for the commercialisation of smallholder production. Hence, taken together, the chapters in Part I provide the theoretical and political context for the subsequent empirical contributions to the book. The latter are divided in three parts according to their analytical focus and methodological framework.

Part II contains four chapters examining different issues pertaining to contract farming, all from within a value chain perspective by adapting concepts from this analytical approach. Hence, the contributions are not conventional global value chain studies in the usual sense of situating the national section of a particular value chain in a global context. Rather, they use one of the key conceptual elements of this perspective (governance, institutional framework, upgrading) as an analytical device with which to study the potential and challenges of a specific contractual arrangement.

Chapter 4 explores the relationships between two main groups of actors, namely smallholders and tobacco-processing companies in the Tanzanian part of the global value chain (GVC) for tobacco. It analyses the contractual system and the relationships between smallholders and tobacco-processing companies in three distinct periods of tobacco contract farming in Tanzania. It argues that changing governance structures are not primarily the result of chain-internal dynamics and that they are not 'driven' by the strategic needs of the lead firms in the GVC for tobacco. Rather, value chain governance is intertwined with the active role of the state in shaping chain structures and socio-economic outcomes.

Chapter 5 focuses on SMEs and issues pertaining to their involvement in contractual arrangements. Some of the vegetable oil-producing SMEs in Tanzania's value chain for sunflower seed have established farming contracts for smallholders in the main sourcing area of Singida and Dodoma Regions. The major problem for these SMEs is to ensure stable and adequate supplies of high-quality sunflower seed, for which contract farming is considered one possible solution. The chapter shows that successful contract farming schemes involving SMEs need to go beyond the sales agreement alone to involve a comprehensive package of additional elements that support smallholder production.

Chapter 6 asks if contract farming facilitates process upgrading, that is, improves the technical and organisational aspects of the production process. The question is examined for smallholder participation in two sugar contract farming schemes in Morogoro Region. The authors examine the incentives for smallholders to acquire the necessary capabilities to improve production processes. The study finds that training, input provisioning and extension services are not included in the contract farming arrangements, despite their crucial role in establishing smallholders' capabilities. Moreover, the incentives included in the arrangements, in the form of the determination of purchase price and payment modalities, are counterproductive as they actually reduce smallholders' incentives to acquire the capability to upgrade the process of sugarcane production.

Chapter 7 marks the final contribution to Part II. This contribution examines the development of the division of labour between two of the main stakeholders in the cotton chain in Tanzania during the pilot phase (2008–11) of contract farming. The authors discuss to what extent improvements in chain coordination have resulted in upgrading of the chain. It is argued that value chain coordination and upgrading appear to be intimately linked in this case, as improvements in coordination have facilitated product and process upgrading. However, the contractual system has been influenced by a number of key constraints and barriers to state-led upgrading interventions in (private) agricultural value chains, to which the chapter also draws attention.

Part III of the book consists of two chapters, both using quantitative methods and cross-sectional data sets collected from Tanzanian farm households. Chapter 8 uses data from tobacco contract farmers in Urambo District and a Logit model to analyse the factors that determine farmers' adoption of input credit. Most farmers (82 per cent) adopt the input credit in tobacco production, whereas cash incomes from other activities (farm or off-farm) and lower degrees of wealth, as indicated by land ownership and type of house-wall construction, increase the likelihood of adopting input credit.

Chapter 9 uses data from smallholder sugarcane out-growers in Kilombero and Turiani and applies Tobit models to analyse the factors that determine households' income diversification. Many sugarcane out-growers receive an income from various (diversified) income sources, whereas the income from these sources and the general diversification depend on many household characteristics.

Part IV consists of chapters that address contract farming in relatively unconventional ways, as the chapters deal with issues like trust, corporate social responsibility and the empowerment of smallholders. These issues are analysed within the framework of particular contract farming arrangements. Chapter 10 examines trust in contract farming arrangements for sunflower production in central Tanzania. Different kinds of trust and how they are established between smallholders and small agro-industrial companies are examined. It is argued that trust plays a significant role in the success and failure of contract arrangements, even though in theory a contract should render the need for trust insignificant. Hence, assessing trust and the nature of personal relations between value chain actors provides valuable insights that can help explain the territorial impact of contract farming arrangements.

Chapter 11 also addresses trust but somewhat conversely, namely by analysing sugarcane out-growers' allegations of dishonest practices against two sugar companies as examples of conspiracy theories. It is argued that these theories express a longing for epistemological certainty (by smallholders) for access to assured knowledge about the weight and sucrose levels of the sugarcane they deliver to the milling companies. By revealing the forged reality of official discourses about sugarcane production they also express a longing for ontological certainty. Hence, the conspiracy theories demonstrate a compelling *will to know* in the way an ideal public sphere would allow. However, for contracted smallholders in a covert sphere, this is severely impeded by forces too strong for them to combat.

Chapter 12 analyses whether two strings of empowerment, the economic and the political, have materialised for smallholders engaged in the contract farming of sunflower seed. It is argued that contract farming through production assistance – particularly improved seed that is accessible only to contract farmers – has increased economic empowerment and resulted in higher economic outcomes compared to non-contracted smallholders. However, the political empowerment that would address the power asymmetries between smallholders and contractors in negotiating contract arrangements is absent. It is concluded that institutional support to assist smallholders to form associations could enhance their bargaining power and thereby become a pivotal means of political empowerment.

Finally, Chapter 13 examines the recent embeddedness of social and environmental concerns in tobacco contract farming schemes in Tanzania. It is argued that the tobacco-processing companies tightly coordinate their contract farming schemes in order to secure crucial on-farm compliance practices and prevent child labour. In addition, by financing local development projects, the ability to allocate resources both among and within communities to individual smallholders complements the tobacco-processing companies' strategic focus on the supply of tobacco leaves while at the same time addressing compliance needs with crucial corporate social responsibility requirements.

References

Bair, J. (2005). Global Capitalism and Commodity Chains: Looking Back, Going Forward. *Competition & Change, 9*(2), 153–180.

Bair, J. (2009). Global Commodity Chains. In J. Bair (ed.), *Frontiers of Commodity Chain Research* (pp. 1–34). Stanford, CA: Stanford University Press.

Barrett, C. B., Bachke, M. E., Bellemare, M. F., Michelson, H. C., Narayanan, S. and Walker, T. F. (2012). Smallholder Participation in Contract Farming: Comparative Evidence from Five Countries. *World Development, 40*(4), 715–730.

Barrientos, S., Gereffi, G. and Rossi, A. (2011). Economic and Social Upgrading in Global Production Networks: A New Paradigm for a Changing World. *International Labour Review, 150*(3–4), 319–340.

Bellemare, M. F. (2015). Contract Farming: What Is in It for Smallholder Farmers in Developing Countries? *Choices*, 30(3), 1–4.

Birthal, P. S., Jha, A. K. and Singh, H. (2007). Linking Farmers to Markets for High Value Agricultural Commodities. *Agricultural Economics Research Review*, 20, 425–439.

Boesen, J. and Mohele, A. T. (1979). *The Success Story of Peasant Tobacco Production in Tanzania: The Political Economy of a Commodity Producing Peasantry*. Uppsala: Scandinavian Institute of African Studies.

Bogetoft, P. and Olesen, H. B. (2004). *Design of Production Contracts: Lessons from Theory and Agriculture*. Copenhagen: Copenhagen Business School Press.

Bolton, P. and Dewatripont, M. (2005). *Contract Theory*. Cambridge, MA and London: MIT Press.

Bryceson, D. F. (1996). Deagrarianization and Rural Employment in Sub-Saharan Africa: A Sectoral Perspective. *World Development, 24*(1), 97–111.

Chambers, R. G. (1988). *Applied Production Analysis: A Dual Approach*. Cambridge: Cambridge University Press.

Chambers, R. G. and Quiggin, J. (2000). *Uncertainty, Production, Choice, and Agency: The State-Contingent Approach*. Cambridge: Cambridge University Press.

Coe, N. M. and Hess, M. (2011). Local and Regional Development: A Global Production Network Approach. In A. Pike, A. Rodríguez-Pose, and J. Tomaney (eds), *Handbook of Local and Regional Development* (pp. 128–138). Abingdon: Routledge.

Coe, N. M., Dicken, P. and Hess, M. (2008). Global Production Network: Realizing the Potential. *Journal of Economic Geography*, 8(3), 271–295.

Diao, X., Hazell, P., Resnick, D. and Thurlow, J. (2007). *The Role of Agriculture in Development: Implications for Sub-Saharan Africa*. IFPRI Research Report No. 153. Washington, DC: International Food Policy Research Institute.

Dicken, P. (2015). *Global Shift*. London: Sage.

Dolan, C. and Humphrey, J. (2004). Changing Governance Patterns in the Trade in Fresh Vegetables between Africa and the United Kingdom. *Environment and Planning A*, 36, 491–509.

Fold, N. (2014). Value Chain Dynamics, Settlement Trajectories and Regional Development. *Regional Studies*, 48(5), 778–790.

Fold, N. and Larsen, M. N. (2008). Key Concepts and Core Issues in Global Value Chain Analysis. In N. Fold and M. N. Larsen (eds), *Globalization and Restructuring of African Commodity Flows* (pp. 26–43). Uppsala: Nordiska Afrikainstitutet.

Gardner, R. (2003). *Games for Business and Economics* (2nd edn). Austin, TX: Leyh Publishing LLC.

Gereffi, G. (1994). The Organization of Buyer-Driven Global Commodity Chains: How U.S. Retailers Shape Overseas Production Networks. In G. Gereffi and M. Korzeniewicz (eds), *Commodity Chains and Global Capitalism* (pp. 95–122). London: Praeger.

Gereffi, G. (1995). Global Production Systems and Third World Development. In B. Stallings (ed.), *Global Change, Regional Response: The New International Context of Development* (pp. 100–142). Cambridge: Cambridge University Press.

Gereffi, G. (1999). International Trade and Industrial Upgrading in the Apparel Commodity Chain. *Journal of International Economics*, 48, 37–70.

Gereffi, G. (2001). Shifting Governance Structures in Global Commodity Chains, with Special Reference to the Internet. *American Behavioral Scientist*, 44(10), 1616–1637.

Gereffi, G., Humphrey, J., Kaplinsky, R. and Sturgeon, T. J. (2001). Introduction: Globalisation, Value Chains, and Development. *IDS Bulletin*, 32(3), 1–8.

Gereffi, G., Humphrey J. and Sturgeon, T. (2005). The Governance of Global Value Chains. *Review of International Political Economy*, 48(1), 78–104.

Gibbon, P. (2008). Governance, Entry barriers, Upgrading: A Re-interpretation of Some GVC Concepts from the Experience of African Clothing Exports. *Competition & Change*, 12(1), 29–48.

Gibbon, P. and Ponte, S. (2005). *Trading down Africa, Value Chains, and the Global Economy*. Philadelphia, PA: Temple University Press.

Hansen, U. E., Fold, N. and Hansen, T. (2016). Upgrading to Lead Firm Position via International Acquisition: Learning from the Global Biomass Power Plant Industry. *Journal of Economic Geography*, 16(1), 131–154.

Hopkins, T. K. and Wallerstein, I. (1994). Commodity Chains: Construct and Research. In G. Gereffi and M. Korzeniewicz (eds), *Commodity Chains and Global Capitalism* (pp. 17–20). London: Praeger.

Humphrey, J. and Schmitz, H. (2002). How Does Insertion in Global Value Chains Affect Upgrading in Industrial Clusters? *Regional Studies*, 36(9), 1017–1027.

Laffont, J. J. and Martimort, D. (2002). *The Theory of Incentives: The Principal Agent Model*. Princeton, NJ: Princeton University Press.

Lawson, V. (2007). *Making Development Geography*. London: Hodder Arnold.

Neilson, J. and Pritchard, B. (2009). *Value Chain Struggles: Institutions and Governance in the Plantation Districts of South India*. Chichester: Wiley-Blackwell.

Ostrom, E. (1998). A Behavioral Approach to the Rational Choice Theory of Collective Action: Presidential Address, American Political Science Association, 1997. *The American Political Science Review*, 92(1), 1–22.

Oya, C. (2012). Contract Farming in Sub-Saharan Africa: A Survey of Approaches, Debates and Issues. *Journal of Agrarian Change*, 12(1), 1–33.

Ponte, S. and Ewert, J. (2009). Which Way Is 'Up' in Upgrading? Trajectories of Change in the Value Chain for South African Wine. *World Development*, 37(10), 1637–1650.

Selwyn, B. (2013). The Global Retail Revolution, Fruiticulture and Economic Development in North-east Brazil. *Review of International Political Economy*, 20(1), 153–179.

Shapiro, S. P. (2005). Agency Theory. *Annual Review of Sociology*, 31, 263–284.

Singh, I., Squire, L. and Strauss, J. (eds) (1986). *Agricultural Household Models: Extensions, Applications, and Policy*. Baltimore, MD: Johns Hopkins University Press.

Sutton, J. and Ulomi, D. (2012). *An Enterprise Map of Tanzania*. International Growth Centre in association with the London Publishing Partnership.

TNBC (2009). *Kilimo Kwanza: Towards Tanzania Green Revolution*. Dar es Salaam: Tanzania National Business Council.

Tokatli, N. (2013). Toward a Better Understanding of the Apparel Industry: A Critique of the Upgrading Literature. *Journal of Economic Geography*, 13, 993–1011.

URT (2016). *National Five Year Development Plan 2016/17–2020/21*. Dar es Salaam: Ministry of Finance and Planning, United Republic of Tanzania.

URT (2010). *Tanzania 1st Five Year Development Plan*. Dar es Salaam: United Republic of Tanzania.

Williamson, O. E. (1991). Comparative Economic Organization: The Analysis of Discrete Structural Alternatives. *Administrative Science Quarterly*, 36(2), 269–296.

Williamson, O. E. (1996). *The Mechanism of Governance*. Oxford: Oxford University Press.

Wu, S. (2006). Contract Theory and Agricultural Policy Analysis: A Discussion and Survey of Recent Developments. *Australian Journal of Agricultural and Resource Economics*, 50, 490–509.

Wu, S. and Roe, B. (2007). Contract Enforcement Social Efficiency and Distribution: Some Experimental Evidence. *American Journal of Agricultural Economics*, 89(2), 243–258.

2 Contract farming

Fluid concept on firm grounds

Lotte Isager, Niels Fold and
Marianne Nylandsted Larsen

Introduction

Contract farming as an agricultural institution has expanded enormously over the past half-a-century and has long been an integral part of farmers' lives in almost every country in the world. Different authors have defined contract farming more or less along the lines of this official FAO definition:

> Contract farming can be defined as agricultural production carried out according to an agreement between a buyer and farmers, which establishes conditions for the production and marketing of a farm product or products. Typically, the farmer agrees to provide agreed quantities of a specific agricultural product. These should meet the quality standards of the purchaser and be supplied at the time determined by the purchaser. In turn, the buyer commits to purchase the product and, in some cases, to support production through, for example, the supply of farm inputs, land preparation and the provision of technical advice.
>
> (FAO, 2015)

To illustrate the popularity of contract farming, a World Bank report (World Bank, 2014) shows that in 2008 contracts accounted for 39 per cent of the total value of the United States' agricultural production, up from 28 per cent in 1991 and 11 per cent in 1969. The *World Investment Report 2009* (UNCTAD, 2009), which pays particular attention to contract farming, reports that 70 per cent of poultry production and 30 per cent of soybean production in Brazil are carried out as contract farming. In Vietnam, 90 per cent of cotton and 40 per cent of rice and tea are grown on a contractual basis, as are and in Pakistan, Nestlé contracts with 140,000 milk-producing farmers. Swinnen and Maertens (2007) reckon that 12 per cent of the rural population in Mozambique is involved in contract farming. In numerous other countries similarly impressive figures could be listed.

There is no doubt that these figures do illustrate a common worldwide tendency towards the integration, through contract farming, of farmers into commercial networks that operate on national, regional or global scales. Although this is a phenomenon that has clearly blossomed since the end of the Second World War, it is actually an institution that has a much longer history in Europe and the United States (Rehber, 2007; Narayanan, 2012). However, contract farming is, quite literally, increasingly gaining ground in developing countries. At the same time, it is important to realise that the statistics refer to hugely diverse and incomparable agricultural practices and organisational forms, even though they all fall under the same rubric of contract farming. Regardless of which definition is employed, contract farming therefore remains a sort of terminological 'black box'. Linguistically, if somewhat misleadingly, the concept is typically referred to in the singular, even though in practice contract farming clearly suggests the plural because it covers such a vast diversity of agrarian experiences. As Carlos Oya notes in his inspirational survey of approaches, debates and issues in African contract farming, the sheer variety of experiences renders it impossible to define a general theory of contract farming due to the 'fluid character of contract farming as an agricultural institution' (Oya, 2012: 17).

In the following, we will elaborate on the empirical diversity of contract farming experiences and discuss some of the methodological and analytical challenges encountered in contract farming research. We illustrate the conceptual diversity of contract farming and attempt to systematise the different combinations of factors in a typology of schemes. We document the recent dominance of econometric studies of contract farming's impact on the incomes of farmer participants and trace this focus to a marked bias towards the household as the key analytical unit. As broader issues of power, inequality, stratification, etc. in communities dominated by contract farming have only been studied thus far to a limited extent, we point to the peculiar relegation of the contract itself – and of interpretations of compliance by the contracting partners – to the sidelines of scholarly research. The conclusion condenses the main argument and findings of the chapter.

The diversity of contract farming

Contract farming is commonly linked to notions of agribusinesses and the vertical coordination of smallholders into global value chains. It is often assumed that contract farming implies a commercial relationship between a private company and a group of farmers, but it is perhaps less widely recognised that the concept also covers commercial relationships between state-managed companies and farmers. Historically, both types of contract-based institutional arrangement have been practised in many African countries. In Tanzania, cotton, sugar and tobacco production were all under state management for several years, a situation that ended with the liberalisation of agricultural markets and the privatisation of parastatals during the 1990s. In addition to companies that are under either full private or public ownership, a third category of companies have a mixture of public and private ownership, an example being the Kilombero Sugar Company, where two private, foreign-based firms own 75 per cent of the shares and the Tanzanian government retains a 25 per cent stake.

Another generally held assumption about contract farming is that it only includes farmers who own the land they cultivate. However, there are numerous cases of contract farmers who grow crops on other people's land either as tenants or share-croppers (Vermeulen and Cotula, 2010) or combine their own land with rented land in the cultivation of crops on a contractual basis.

Apart from the diversity of land tenure and the variable ownership structures inherent in contract farming relations, the fluidity of the concept of contract farming is largely due to the fact that it covers a vast multitude of organisational arrangements. To illustrate, the contract farming of tobacco in Mozambique involves just one company (Mozambique Leaf Tobacco, the local subsidiary of Universal Leaf Tobacco), which holds a national monopsony and maintains contracts with around 130,000 individual growers (Pérez-Niño, 2014). In neighbouring Tanzania, the contract farming of tobacco is structured completely differently. Here, two processing companies – sometimes acting individually and at other times acting jointly in the form of the Association of Tanzanian Tobacco Traders – have contractual relationships with a number of so-called primary cooperative societies, whose members are individual tobacco farmers. The primary cooperative societies are organised in the Regional Cooperative Union and the Tanzania Cooperative Apex, which, along with the Tanzania Tobacco Council, representing the main stakeholders in the sector, all form part of how the tobacco schemes are organised.

The degree of organisational complexity in a contracting scheme may depend on several factors, ranging from biological crop characteristics to the political history of a given country or region. Crops such as sugarcane, oil palm, fruits and vegetables, which are highly perishable, require stringent and efficient processing and marketing procedures in order to retain their market value. For these crops, their very nature implies specific logistical demands, which to a large extent explains why the production is organised in certain ways. But for other crops such as tobacco, mentioned above, the differences in contractual arrangements cannot be explained purely by the nature of the crop but rather by different social and political traditions in different countries or regions.

One way of establishing a generic overview of the diverse forms of contract farming schemes is Shepherd's typology (2013), which contains five basic organisational models: (1) *centralised*, where the company processes the crop and needs delivery from growers of large volumes of persistent quality. The company is very likely to hold a monopsony; (2) *nucleus estate*, a variant of centralisation, where a company's estate production is supplemented by crops from surrounding growers; (3) *multipartite*, a joint venture where private and public partners deliver inputs, extension services and other services to growers; (4) *informal*, a set-up that may simply include a marketing contract, with other services added or not, between growers and a company; and (5) *intermediary*, where one or several private companies have contractual relationships with a producer organisation or cooperative, a growers' union, a village or district committee or other parties.

Following this typology, tobacco production in Mozambique under monopsony would be categorised as a centralised organisational model, whereas tobacco production in Tanzania exemplifies an intermediary model. Hence categories like these make it possible to carry out systematic comparisons between similar organisational forms or to consciously compare different forms, perhaps of contract farming schemes in similar crops. In addition, trajectories of different organisational forms can be analysed and the reasons for change explained.

The impact of contract farming: income benefits or welfare?

One of the most central questions about contract farming is naturally whether participant farmers benefit from it or not. Researchers have drawn highly discrepant conclusions about this issue. We shall not repeat the conclusions here but instead discuss the at times paradoxical nature of the debate by dwelling on the notion of household income. Treated as a credible indicator of the status of general household welfare, household income has continually been a key concern for contract farming researchers. In *Living under Contract* (Little and Watts, 1994), one of the classic publications on contract farming, Michael Watts (1994) refers to studies from the 1980s and early 1990s that show a fairly consistent positive correlation between participation in contract farming and increasing household income. This trend has been substantiated by several studies conducted during the past decade, which are based on a more solid econometric methodology than the older studies (see Bellemare, 2010, 2012; Maertens and Swinnen, 2009; Minten *et al.*, 2009; Miyata *et al.*, 2009; Guimas, 2012; Junichi *et al.*, 2012).

Based on these studies, it would presumably be straightforward to conclude that contract farming leads to higher income levels for the participant households. The available data point in that direction, and influential publications such as the *World Investment Report 2009* draw on these data to promote the further expansion of contract farming in developing countries (UNCTAD, 2009). However, it is important to keep in mind – as Watts (1994) does, but more recent studies tend to ignore – that higher household income might not automatically translate into more general and long-term welfare benefits such as better nutritional and health status in the household as a whole, higher educational levels for members of the household (including female household members) or new investments in housing, agriculture or other income-generating activities. A study by Glover and Kusterer (1990) found that contracted farmers with a surplus tended to invest in education and household capital but that this in itself had not created a 'knock-on effect' on their general livelihoods. Ten years after this study, in a literature review that included Glover and Kusterer (1990) as reference, Baumann (2000) examined correlations between household income and nutrition but found no 'spillover effects'. The status of this question may therefore be summed up by saying that, even though the positive relationship between participation in contract farming and increasing household income has now been firmly documented, no study of contractual schemes has yet been able to prove the long-term effects of contract farming on social welfare in the wider sense of the word.

Barrett *et al.* (2012) add a notable and thought-provoking argument to the ongoing discussion about the economic benefits of contract farming in terms of increasing household incomes by stressing that the significance of the procurement location of firms' contracting choices is routinely overlooked by researchers. This means that researchers are liable to conclude that particular contract farming arrangements are successful because of the way they are organised, even though it would be more accurate to explain their success simply by their choice of location.

With regard to the discussion of the economic benefits of contract farming, it is also noticeable that even neoclassical economists acknowledge that contract farming works best 'when the contracting firm is explicitly granted a monopsony, as this helps to enforce contracts, prevent side-selling and thus ensure the success of the operation' (Grosh, 1994; Kirsten and Sartorius, 2002; World Bank, 2011: 70, all cited in Smalley, 2013). Hence, it is paradoxical that a form of agricultural production that is supposed to help integrate farmers into large-scale economic networks based on free-market principles is best suited to monopsonist settings in which these principles must be suspended in order to function adequately (Smalley, 2013; Oya, 2012).

Assessing contract farming: household scale or holistic approaches?

Considering the great variety of ways in which contract farming schemes are organised in different countries, it is curious that the contract farming literature published over the past 10–15 years is overwhelmingly dominated by economic case studies in which the unit of analysis is the household and the investigation is limited to fairly simple contractual arrangements. This analytical focus on households is an important explanation for the lack of attention to broader issues within the political economy of contract farming. Moreover, the timescale of most studies is restricted to covering either a single production season or, alternatively, a two- to three-year period. This means that the long-term consequences of contract farming schemes, whether economic, social or environmental, remain unrecognised, or at least untold, in most publications (but see Richardson, 2010).

Even in cases where other parties, such as banks or credit institutions, government agencies or NGOs, are clearly important for the functioning of the contractual scheme as a whole, the contract farming literature appears to be chiefly concerned with just one set of relationships – firm versus farmer/household head. The contemporary literature is thus characterised by a turning away from the more holistic approach found in earlier contract farming studies, such as Steven M. Jaffee's analysis of contract farming in Kenyan horticulture (1994). Jaffee argued that contract farming schemes should be explored within their respective historical contexts and that 'attention should focus not simply on the relationship between "the firm" and the farmers', since other relationships – for example, between company management and cooperative officials – might be equally important (Jaffee, 1994: 136).

Arguably, Jaffee's appeal for holism has largely been ignored in much subsequent research. For example, 'little research has been done on the role played in contract farming by farmer organisations' (Shepherd, 2013: 7). To the extent that these organisations are mentioned at all, it is usually without providing many empirical details of their history and role in a contractual system, and the purpose of mentioning them is generally to extract brief 'lessons learned' about very particular aspects of the performance (e.g. Kirsten and Sartorius, 2002; Coulter *et al.*, 1999; Bijman, 2008; Sivramkrishna and Jyotishi, 2008; Glover, 1987; Bingen *et al.*, 2003). The literature review by Baumann (2000) is characteristic of this trend in stressing particular lessons learned about the function of most farmers' organisations that have emerged within contract farming schemes. These lessons are, first, that they serve to regulate production, and second, that they do not constitute a forum within which small farmers can fight for better conditions or greater control over management. In principle, Baumann concludes, farmers' organisations and cooperatives should 'be the ideal institution for integrating all stages of the production process and ensuring farmers are fully represented. However, the experience of cooperatives, mainly due to poor farmer representation, bureaucratic inefficiency and political manipulation, has not been very good' (Baumann, 2000: 17).

In an attempt to examine the role farmers' organisations play in contract farming schemes in greater empirical detail, Isager *et al.* (in press) have analysed the post-privatisation roles of out-growers' associations in the contract farming of sugarcane in Tanzania. They conclude that the privatisation process has transformed the associations' leadership into powerful entrepreneurs and mediators of private capital accumulation. As examples of emergent local capital, the associations should not be portrayed simply in terms of the negative stereotypes – corrupt, non-transparent, unaccountable – that are commonly regurgitated in other papers. Rather, they need to be understood as actors that add a significant dimension to regional dynamics while being positioned within the contemporary alliances between national, regional or global capital in which contract farming in Africa takes place.

In light of Jaffee's arguments in favour of maintaining a holistic approach to contract farming, it is worth contemplating more elaborately some of the effects of an exclusive emphasis on the household that is found in many recent contract farming studies. In order to do so it is illustrative to look at two exemplary papers, which both deal with the French company Lecofruit's contract farming schemes in Madagascar (Bellemare, 2010; Minten *et al.*,

2009). Both papers explain that Lecofruit has organised production by hiring 300 local team leaders ('*chefs de culture*'), each of whom coordinates five to six technical assistants ('*assistants de culture*') who supervise people from their own village. This 'allows the processor to exploit local knowledge of the growers it contracts' (Bellemare, 2010: 508) and ensures 'correct production management as well as avoid[ing] side-selling' (Minten *et al.*, 2009: 1733). Presumably, these *chefs* and *assistants* are high-status members of their local communities, and the company can therefore draw upon their existing social capital and authority to exercise power and production control and, ultimately, secure an economic capital gain. However, neither paper explains who these *chefs* and *assistants* are, what their social and political statuses are and the criteria according to which they were selected. It is therefore difficult to understand how Lecofruit's way of organising its production might influence, and be influenced by, the social and political lives of either the growers or the non-growers.

This is not a trivial matter. Intuitively, the findings reported in these two papers sound like very good news. Contracting with Lecofruit has brought increasing incomes to hundreds of Madagascan households (Bellemare, 2010; Minten *et al.*, 2009). The scheme could therefore be hailed as a success, as both papers conclude. But does the 'success' in increasing incomes not depend also on whether Lecofruit's engagement in Madagascar is underwriting a hierarchical political system that may exclude major sections of the rural population from taking part in political and economic decision-making? The lack of attention to such issues due to the household bias in both papers has the consequence that the positive results reported in the two econometrically solid studies remain opaque with regard to the deeper value of increased incomes in the lives of Madagascan farming households and in the rural transformation processes they are part of. We contend that the great majority of contract farming studies that seek to document the income benefits of contract farming for households suffer from the same lack of attention to the wider social and cultural effects of contract farming.

That said, it is relevant to pay attention to some of the few exceptions to the general trend of studying contract farming purely from the perspective of whether this form of production leads to higher household incomes or not. Andrew Walker's (2009) analysis of a group of Thai farmers' adoption of contract farming is a significant case in point. Walker considers how the contract farming of garlic in a northern Thai village is embedded in local systems of sociality and livelihood diversification. This enables him to capture the complexity of firm–farmer roles and relationships, including the degree of leverage farmers have due to other and perhaps more important sources of income than agriculture. Besides providing a far more complex and empirically rich account of the relationship between firms and farmers than is normally offered in studies of contract farming, the scenario presented by Walker is of interest for other reasons too. In contrast to the pervasive tendency of agribusinesses in many countries to choose geographical locations for contracting that are characterised by adequate infrastructure and minimal competition from other firms (cf. Barrett *et al.*, 2012), the situation outlined by Walker is somewhat different.

> There is no doubt that as individual economic agents the companies are much more powerful than the farmers. As accounts of corporate proletarisation rightly point out, the seemingly reciprocal contracts gloss over substantial disparities in economic resources and socio-political influence; but relationships of power are not simply determined by structural position. In the everyday exercise of power, contracting companies have to operate in an environment where there is considerable competition for the land and labour of farmers. A vibrant national and international agricommodity trade means that the companies are keen to secure contracts with farmers who own land in suitable agro-ecological zones. As rural households increasingly adopt diverse livelihood strategies, companies recognise that agriculture is not the sole, or even the most important, source of local income. In this climate of economic diversity and choice, companies cannot simply impose their will. They have to be careful to fit within locally valued systems of economy and sociality.
>
> (Walker, 2009: 69)

In this northern Thai scenario, location is still all-important for the contracting companies. But, unlike numerous African locations with relatively little competition over access to farmers' land and labour, firms must compete hard in Thailand. Considering the ongoing expansion of contract farming into developing countries, it is quite possible that the situation Walker describes for Thailand points to a future scenario that could become a reality in comparatively remoter locations (e.g. in Africa) that are currently in the process of being drawn into the world market in the same way as Thai farmers have been for several decades.

In addition to Walker's study of garlic farmers in Thailand, Finnis's research (2006) on the multi-faceted social process of the adaptation of contract farming in the Kolli Hills in India also breaks the general pattern of household-biased studies of contract farming. By investigating a long-standing communal negotiation of both household and community goals, Finnis convincingly demonstrates how households are embedded in wider social and cultural structures and practices and how the different scales of analysis mutually affect each other.

Another non-household-focused study, also from India, by Pritchard and Connell (2011) shows how the implementation of contract farming is associated with the strategies buyers follow to mitigate risk and uncertainty. This study suggests, in line with the econometric studies cited earlier, that contract farming tends to benefit participating farmer households. However, this finding is put into a larger perspective by discussing the value of monetary benefits in the context of the environmental and social implications of contract farming, which remain contested in India. Much in line with other observers of contract farming (e.g. Oya, 2012; Eaton and Shepherd, 2001), Pritchard and Connell (2011) conclude that, because the global market is highly uneven and unpredictable, the structural nature of contract farming and the social relations of its enforcement differ from place to place and from product to product. Hence, Pritchard and Connell (like Walker and Finnis mentioned above) help move the analysis of contract farming beyond the household scale by emphasising the ways in which contract farming schemes spatially integrate agrarian landscapes into wider economies while simultaneously reproducing spatial differentiation because the nuances of schemes are necessarily adaptive to the institutional environments in which they are grounded.

Contracts and enforcement

It is a peculiar paradox of the contract farming literature that the only link that unites the enormously diverse empirical cases of contract farming in the world is the existence of some sort of contract – and yet the contract is the least empirically studied or theoretically discussed aspect of contract farming. It is very rare to find a contribution that provides information about the theoretical understanding of contracts on which the paper is based. Instead, it is implicit in most papers that certain qualities and attributes are ascribed to the contract, but these characteristics are taken for granted rather than explicitly noted and discussed. In order to elaborate further on this issue it is useful to borrow the contract typology drawn up by Suchman (2003), who is not a contract farming researcher but a legal scholar. Suchman (2003) distinguishes between three basic understandings of contracts:

- *contract-as-doctrine*: This understanding of the contract is concerned with how and why actors acquire formal rights. It is assumed that contracts are binding agreements entered into by two or more equal parties. The analytical focus is on formal obligations and procedures and official principles for conducting business and collaboration.
- *contract-as-relation*: The underlying premise of this understanding of contracts is that it questions why and how transacting parties actually secure each other's performance. It is assumed that legal doctrine and legal recourse usually matter very little in concrete transactions because in practice they are governed by informal community norms and enforced by informal social sanctions.
- *contract-as-artefact*: This understanding of contracts addresses the fundamental question of why so many people bother to spend lots of resources on making quite elaborate, formal and legally binding contracts (including sections with procedures for dispute resolution) that none of the parties will ever use in practice. Why and how do actors produce various types of contract document? Why and how do particular styles, forms

> and features enter and leave the contractual repertoire over time? Hence, this understanding stresses that a contract as an artefact may have cultural meanings that are unrelated to their legal content.

What is emblematic of contract farming papers is that they inevitably proceed without specifying which understanding(s) of the contract are employed in the analysis. If a contract is described at all, it is usually very brief and with a focus on technical terms such as the duration of the contract and the basic conditions of exchange between companies and farmers. That is like asking whether the formal obligations and procedures for collaborating and conducting business are absolutely crucial, but not asking whether they are relevant for an understanding of how the contractual scheme functions in practice, nor, if so, to what extent.

Alternatively, the contract is mentioned indirectly because one of the parties does not abide by it but instead engages in 'morally hazardous' behaviour such as side-selling, cheating or corruption – which are allegedly epidemic in many contract farming arrangements in developing countries. To the extent that such behaviour is analysed, it is inevitably and quite logically dealt with as a matter of 'enforcement'. Some papers adopt a 'contract-as-relation' understanding and pay empirical attention to the relatively informal enforcement mechanisms employed by companies that operate in environments with ineffective public enforcement institutions. The papers by Bellemare (2010, 2012) and Minten *et al.* (2009) mentioned earlier, which analyse Lecofruit's collaboration with local authority figures in Madagascar, are examples of this.

However, most papers stress the need for more formalised contracts and effective government enforcement through the creation of appropriate legal, judicial and institutional environments (e.g. World Bank, 2014). White (1997), for one, contends that governments and NGOs simply need to put the right policies in place and establish proper legal frameworks to protect farmers from exploitation by companies and, vice versa, relieve companies from problems with side-selling, etc. The underlying premise of all this advice is a 'contract-as-doctrine' approach that reproduces a highly instrumentalist concept of the contract by arguing that clear and cogent contracts are the main preconditions for successful contract farming.

It is important to emphasise the plentiful empirical evidence from developing countries showing that even the most detailed and elaborate contracts tend to be far less effective in practice than suggested by authors subscribing to the 'contract-as-doctrine' perspective. Narayanan (2012) and Barrett *et al.* (2012) have explored enforcement in theoretically and empirically very useful ways (see also Bijman, 2008). In her analysis of contract farming in India, Narayanan (2012) argues convincingly that firms in fact prefer fairly unelaborated contracts because they allow them to build up stronger relationships with farmers, resulting in more productive farming results in the long run. She further criticises the tendency in the contract farming literature to describe particular cases in categorical terms as being either formal (written contracts) or informal (oral agreements), even though real-life interaction and collaboration between contracting parties are often more mixed and complex (Narayanan, 2012: 22). This finding is no doubt valuable to keep in mind for policy-makers, civil servants, NGOs and others in Africa who are striving hard to improve the formal quality and content of contracts.

Finally, it is worth stressing that few studies of contract farming dwell at any length on the issue of whether contracted farmers actually know and understand their contracts. Some papers note briefly that 'asymmetric information' is found in a given contractual system, but the topic is usually not given any analytical attention beyond that. Nonetheless, again it is not a trivial matter if people understand the conditions and rules of their mutual work and business relationships. Indeed, living under contracts and conditions that are unintelligible or inaccessible to (some) people may affect their world views and sense of existential certainty and security in profound ways (see Chapter 11, this volume). By relegating such issues to simply being questions of 'asymmetric information' or the older but equally murky concept of 'the mystification of the contract' (Clapp, 1994), the contract farming literature tends to trivialise an essential cultural aspect of many contractual arrangements.

It is possible that this trivialisation is related to the pervasive tendency in the contract farming literature to deal primarily with this agricultural model within a framework of economic (or more precisely econometric) theory. We would argue that this perspective tends to ignore the possibility that contracts may have other important features than those included under the 'contract-as-doctrine' and 'contract-as-relations' perspectives. However, as kinds of social artefacts, contracts may also symbolise social categories and influence social actions (cf. Suchman, 2003), thereby helping to build up social identities and create a sense of loyalty or disloyalty between the contracting parties. Finnis's (2006) analysis of an Indian farmer community's decision-making process at a time when they had to decide if they wanted to engage in the contract farming of tapioca instead of growing their traditional subsistence crops is a fine example of a 'contract-as-artefact' perspective. By treating contract farming as a processual phenomenon as opposed to a more or less reified set of relationships and by stressing the symbolic significance of contract farming that conveys a feeling of being 'more civilised' to the group, Finnis demonstrates how contracts may not only be created by organisations but, conversely, create organisation as well.

Conclusion

Contract farming is increasingly being introduced in rural communities where access to land for large-scale production is scarce and/or restricted due to the local political economy. Despite its huge diversity in terms of ownership, tenure, crop characteristics and organisational forms, the concept and phenomenon are widely hailed as a solution to the limited commercialisation of smallholder production in Africa by national governments, international donors and private (transnational) capital. Recent research findings have confirmed that, generally speaking, contract farmers are experiencing increased incomes, although the broader coupling with better welfare is somewhat more dubious. Hence, whether the increase in incomes translates into better nutritional status, better education, more civic rights, etc. for the whole household is not clear. Likewise, the relationship between contract farming and broader societal trajectories – that is, the consequences of contract farming on local or regional socio-economic dynamics in areas dominated by contract farming – is only sparsely illuminated.

This chapter has argued that the reason for the lack of attention to broader issues connected with contract farming is a shift in scholarly interest from an initial concern with exploitation and unequal relationships between farmers and purchasing companies (in some cases also processing companies) to a focus on the impact on household income from participation in contract farming. In other words, in disciplinary terms there has been a shift from sociology, political science and human geography to economics, in particular advanced econometric analysis. This recent bias towards 'quantitative studies' implies that 'qualitative studies' of the impact of contract farming on local hierarchical structures, accumulation 'on the ground' and social stratification needs to be revitalised and strengthened by systematically incorporating these issues into development studies. This is not to say that the pendulum needs to swing back to its original starting point but rather to emphasise the importance of both types of studies, as well as the potential interaction or alignment of the two general methodologies. Of crucial interest for a more holistic analytical approach to contract farming is the neglected but straightforward issue of the contract itself – how it is understood by the contracting parties (if it is understood at all) and how the issue of contract compliance should be handled. These questions are not only interesting in a scholarly context but of the utmost importance to understand if the aim is to design equal, fair, efficient and sustainable contract farming schemes.

References

Barrett, C. B., Bachke, M. E., Bellemare, M. F., Michelson, H. C., Narayanan, S. and Walker, T. F. (2012). Smallholder Participation in Contract Farming: Comparative Evidence from Five Countries. *World Development, 40*(4), 715–730.

Baumann, P. (2000). *Equity and Efficiency in Contract Farming Schemes: The Experience of Agricultural Tree Crops.* Working Paper 139, Overseas Development Institute, London.

Bellemare, M. (2010). Agricultural Extension and Imperfect Supervision in Contract Farming: Evidence from Madagascar. *Agricultural Economics, 41*, 507–517.

Bellemare, M. (2012). As You Sow, So Shall You Reap: The Welfare Impacts of Contract Farming. *World Development*, 40(7), 1418–1434.

Bijman, J. (2008). *Contract Farming in Developing Countries: An Overview.* Working Paper, Wageningen University.

Bingen, J., Serrano, A. and Howard, J. (2003). Linking Farmers to Markets: Different Approaches to Human Capital Development. *Food Policy*, 28, 405–419.

Clapp, R. A. (1994). The Moral Economy of the Contract. In P. D. Little and M. J. Watts (eds), *Living under Contract: Contract Farming and Agrarian Transformation in Sub-Saharan Africa* (pp. 97–139). Madison: University of Wisconsin Press.

Coulter, J., Goodland, A., Tallontire, A. and Stringfellow, R. (1999). *Marrying Farmer Cooperation and Contract Farming for Service Provision in a Liberalising Sub-Saharan Africa.* Working Paper, ODI Natural Resource Perspectives, 48, November.

Eaton, C. and Shepherd, A. W. (2001). *Contract Farming: Partnerships for Growth.* FAO Agricultural Services Bulletin 145. Rome: Food and Agriculture Organization of the United Nations.

FAO (2015). Contract Farming Resource Centre. Online: www.fao.org/ag/ags/contract-farming/faq/en. Accessed 16 September 2015.

Finnis, E. (2006). Why Grow Cash Crops? Subsistence Farming and Crop Commercialization in the Kolli Hills, South India. *American Anthropologist*, 108(2), 363–369.

Glover, D. J. (1987). Increasing the Benefits to Smallholders from Contract Farming: Problems for Farmers' Organizations and Policymakers. *World Development*, 15(4), 441–448.

Glover, D. and Kusterer, K. (1990). *Small Farmers, Big Business: Contract Farming and Rural Development.* Basingstoke: Macmillan.

Guimas, C. (2012). Effect of Organic Contract Farming on Labor Demand: A Study Case in the Western Uganda. *Economies and Finances.* dumas-00802135.

Isager, L., Fold, N. and Nsindagi, T. (in press). The Post-Privatization Role of Outgrowers' Associations in Rural Capital Accumulation: Contract Farming of Sugarcane in Kilombero, Tanzania. *Journal of Agrarian Change.*

Jaffee, S. M. (1994). Contract Farming in the Shadow of Competitive Markets: The Experience of Kenyan Horticulture. In P. D. Little and M. J. Watts (eds), *Living under Contract: Contract Farming and Agrarian Transformation in Sub-Saharan Africa* (pp. 97–139). Madison: University of Wisconsin Press.

Junichi, I., Bao, Z. and Su, Q. (2012). Distributional Effects of Agricultural Cooperatives in China: Exclusion of Smallholders and Potential Gains on Participation. *Food Policy*, 376, 700–709.

Kirsten, J. and Sartorius, K. (2002). Linking Agribusiness and Small-scale Farmers in Developing Countries: Is There a New Role for Contract Farming? *Development Southern Africa*, 19(4), 503–529.

Little, P. D. and Watts, M. J. (eds) (1994). *Living under Contract: Contract Farming and Agrarian Transformation in Sub-Saharan Africa.* Madison: University of Wisconsin Press.

Maertens, M. and Swinnen, J. F. M. (2009). Trade, Standards, and Poverty: Evidence from Senegal. *World Development*, 37(1), 161–178.

Minten, B., Randrianarison, L. and Swinnen, J. F. M. (2009). Global Retail Chains and Poor Farmers: Evidence from Madagascar. *World Development*, 37(11), 1728–1741.

Miyata, S., Minot, N. and Hu, D. (2009). Impact of Contract Farming on Income: Linking Small Farmers, Packets, and Supermarkets in China. *World Development*, 37(11), 1781–1790.

Narayanan, S. (2012). *Notional Contracts: The Moral Economy of Contract Farming Arrangements in India*, Indira Gandhi Institute of Development Research, Mumbai, September. Online: www.igidr.ac.in/pdf/publication/WP-2012-020.pdf. Accessed October 2013.

Oya, C. (2012). Contract Farming in Sub-Saharan Africa: A Survey of Approaches, Debates and Issues. *Journal of Agrarian Change*, 12(1), 1–33.

Pérez-Niño, H. (2014). O Crescimento da Cultura do Tabaco em Moçambique: Questões e Desafios para um Sector em Transformação. *Desafios para Moçambique*, 99–134.

Pritchard, B. and Connell, J. (2011). Contract Farming and the Remaking of Agrarian Landscapes: Insights from South India's Chilli Belt. *Singapore Journal of Tropical Geography*, 32, 236–252.

Rehber, E. (2007). *Contract Farming: Theory and Practice*. Hyderabad: ICFAI University Press.

Richardson, B. (2010). Big Sugar in Southern Africa: Rural Development and the Perverted Potential of Sugar/Ethanol Exports. *Journal of Peasant Studies*, 37(4), 917–938.

Shepherd, A. W. (2013). *An Introduction to Contract Farming*. Technical Centre for Agricultural and Rural Cooperation (ACP-EU). Online: http://makingtheconnection.cta.int/sites/default/files/Contract-Farming-Introduction.pdf. Accessed 11 August 2015.

Sivramkrishna, S. and Jyotishi, A. (2008). Monopsonistic Exploitation in Contract Farming: Articulating a Strategy for Grower Cooperation. *Journal of International Development*, 20(2), 280–296.

Smalley, R. (2013). *Plantations, Contract Farming and Commercial Farming Areas in Africa: A Comparative Review*. Working Paper 055, Land and Agricultural Commercialisation in Africa (LACA) project Working Paper Series, Institute for Poverty, Land and Agrarian Studies (PLAAS), University of the Western Cape, South Africa.

Suchman, M. C. (2003). Contracts as Social Artifacts. *Law & Society Review*, 37(1).

Swinnen, J. and Maertens, M. (2007). Globalization, Privatization, and Vertical Coordination in Food Value Chains in Developing and Transition Countries. *Agricultural Economics*, 37(1), 89–102.

UNCTAD (2009). *World Investment Report 2009: Transnational Corporations, Agricultural Production and Development*. New York and Geneva: United Nation Conference on Trade and Development (UNCTAD).

Vermeulen, S. and Cotula, L. (2010). *Making the Most of Agricultural Investment: A Survey of Business Models That Provide Opportunities for Smallholders*. London: IIED; Rome: FAO; Rome: IFAD; Bern: SDC.

Walker, A. (2009). Now the Companies Have Come: Local Values and Contract Farming in Northern Thailand. In D. Caouette and S. Turner (eds), *Agrarian Angst and Rural Resistance in Contemporary Southeast Asia* (Chapter 4). Abingdon: Routledge.

Watts, M. J. (1994). Life under Contract. In P. D. Little and M. J. Watts (eds), *Living under Contract: Contract Farming and Agrarian Transformation in Sub-Saharan Africa* (pp. 21–77). Madison: University of Wisconsin Press.

White, B. (1997). Agroindustry and Contract Farmers in upland West Java. *Journal of Peasant Studies*, 24(3), 100–136.

World Bank (2014). *An Analytical Toolkit for Support to Contract Farming*. Washington, DC: International Bank for Reconstruction and Development/World Bank.

3 Strategies and policies for agricultural development in Tanzania

Joseph Andrew Kuzilwa, Daniel Mpeta, Marianne Nylandsted Larsen and Niels Fold

Introduction

This chapter presents an overview of continuity and change in the agricultural strategies and policy mechanisms that have been adopted by the Tanzanian state to develop the agricultural sector. The aim is to provide a policy context for the issues covered in the subsequent chapters of this book. The different 'milestones' in the policy and strategy frameworks are divided into three relatively distinct periods tracing the transition from a state-led to a liberalised economy and the interventions that have been undertaken to transform the sector. Of particular interest are the strategies that were pursued during the post-liberalisation period, when markets were allowed to work and the private sector was given a dominant role in addressing some of the major challenges. This includes the low productivity and inadequate market access for smallholders, chronic problems that have hindered modernisation of the sector. In the most recent period, contract farming has been incorporated into policy documents as an organisational device to 'fix' these bottlenecks and barriers in smallholder-dominated agricultural value chains. In order to position these policies in a material context, the chapter briefly introduces the role of the agricultural sector in industrial development in Tanzania and the importance of the four agro-industrial crops that form the empirical basis for this book. Further details of the production and contractual arrangements are given in the individual chapters.

The role of cotton, sunflower seed, tobacco and sugarcane as agro-industrial crops

It has been a major aspiration of all Tanzanian governments since independence to develop a modern agricultural sector characterised by high productivity and reasonably high incomes for the rural population. The agricultural sector currently contributes 30 per cent of total exports (mainly in the form of commodities) and about 65 per cent of the raw material supply for domestic industries. It also accounts for about 23 per cent of GDP and employs nearly 67 per cent of the national workforce, a far greater share than the industry and service sectors, which jointly employ just 25 per cent of the workforce (URT, 2016a). Despite its importance, the agricultural sector in Tanzania has continually registered lower growth than other sectors, thus affecting its contribution to poverty reduction in the country (UNDP/URT, 2015). For example, while agriculture grew by an average of 4 per cent between 1998 and 2009, during the same period the industry and service sectors increased by averages of about 8 per cent and 7 per cent, respectively (FAO, 2013). During 2010–15 the sector grew at 2.3 per cent, well below the 6 per cent target set in the National Five Year Development Plan for 2010/11–2015/16 and the National Strategy for Growth and Reduction of Poverty II for 2010–15 (URT, 2016a). The main reasons for the modest growth trajectory of the sector include low productivity due to inadequate agricultural financing, poor production techniques and underdeveloped markets, infrastructure and farm-level added value (see, for instance, URT, 2011a).

Due to its strong forward and backward links with the manufacturing sector, the agricultural sector also has a strong potential to support the industrialisation process of the country through value-addition activities in food- and fibre-related industries. All of the four crops (tobacco, cotton, sunflower and sugarcane) that are examined in this book are of special significance in this respect because of their contribution to national and local economic development in the areas where they are grown. In Tanzania, the four crops are predominantly produced by smallholders, and they all feature as priority crops in the first and second National Five Year Development Plans covering the periods 2010/11–2015/16 and 2016/17–2021/21 (URT, 2011b, 2016a).

Tobacco was introduced in Tanzania in 1951, and it is now an important crop for the national economy. Currently it supports the livelihoods of more than 100,000 households that grow it, and it generates substantial employment in the trading and transportation of leaf tobacco and the processing of dried tobacco before it is exported. More than 51 per cent of the crop is grown in western Tanzania (Urambo, Sikonge and Uyui Districts). As shown in Table 3.1, there has been a sharp increase in the production of tobacco, from 55,000 metric tonnes in 2008 to an average of 105,000 metric tonnes in

Table 3.1 Cash crops production trend: 2008–14 (metric tonnes)

Crop	*Year of production*						
	2008	*2009*	*2010*	*2011*	*2012*	*2013*	*2014*
Cotton	200,662	267,004	163,644	250,000	351,151	245,831	203,313
Tobacco	55,356	60,900	130,000	123,904	74,240	100,000	105,881
Sugar	276,605	279,850	317,000	304,135	286,380	294,300	304,007
Sunflower	418,317	466,831	348,877	786,902	1,125,000	2,625,000	2,755,000

Source: URT (2011, 2013, 2014, 2015); Economic Surveys (2011, 2013); Government Budget speeches (2014, 2015) by Ministry of Agriculture, Livestock and Fisheries.

2014, mainly due to an attractive price which persuades farmers to produce more tobacco (URT, 2015), proper use of inputs supported by input credits and good rainfall distribution in the tobacco-growing areas. Between 2014 and 2016, tobacco earned more foreign exchange for the country than any other traditional agricultural product (Bank of Tanzania, 2016). Despite international efforts to control tobacco production (WHO, 2003), the crop has become Tanzania's most successful agricultural product grown specifically for export (Bank of Tanzania, 2016). The crop also contributes to the revenue of the local councils through a crop cess. The crop has been grown under different forms of contract since its initial establishment.

Cotton is grown by between 350,000 and 500,000 smallholder farmers, mainly in the north-western regions of the country – Simiyu (46 per cent), Shinyanga (15 per cent), Mwanza (10 per cent), Geita (12 per cent) Mara (8 per cent), Tabora (4 per cent) and Kagera (2 per cent) – that together constitute the Western Cotton Growing Zone (URT, 2016a). Despite the low absorption capacity of cotton by domestic industry, cotton supports employment for people involved in the lint, textile and garment industries, estimated to be around 13,000 in 2015 (URT, 2016b). In the past decade, cotton has ranked as the second or third (competing with coffee) largest export crop for Tanzania (FAO, 2013). Since 2008, the crop has experienced mixed performance, with production fluctuating over the period: the peak output of 351,000 tons recorded in 2012 was mainly due to attractive world market prices, but production declined to 203,300 tons in 2014. The fluctuating performance has been attributed to declining world market prices that affect farm gate prices, climate changes affecting rainfall levels, the lesser ability of smallholders to buy inputs and farmers not adhering to recommended agronomic practices (URT, 2015). Contract farming arrangements in cotton production is of recent origin in the post-liberalisation era, having been introduced in the Western Cotton Growing Zone in late 2000s by the Tanzania Cotton Board. Under this arrangement, farmers enter into contracts with ginneries, which provide farm inputs on credit and recover input expenses after the sale of cotton lint.

Sugarcane production was introduced in the country in the early 1960s to support the sugar factories that had been established in Moshi, Morogoro and subsequently Kagera. Sugarcane is grown in large-scale nucleus estates but supplemented by out-growers, who produce about 60 per cent of the required cane. Kilombero and Turiani in the east of the country (Morogoro Region) are among the major production areas. The importance of sugarcane production arises from the unsatisfied national demand for sugar, estimated in 2014 at 420,000 tons, of which 170,000 tons are for industrial use (URT, 2016a). Under current demand conditions, the country has to import about 290,000 tons of sugar to make up for the production deficit. As shown in Table 3.1, sugar production is characterised by an increasing trend – even though fluctuating over the period – with production reaching 304,000 tons in 2014. The main reason contributing to the improved smallholder production performance of sugarcane has been use of inputs supported by access to credit and the price offered for sugarcane (which depends on the sugar content). Contracts are signed between the milling companies, which control both the factories and the nucleus estates, and associations of smallholders, the latter being responsible for organising both supplies of input and the harvesting and transportation of sugarcane to the factories.

Sunflower is grown mainly in the central part of the country, but it is also spreading to the north and south. It has of late changed from being of minor commercial importance to being a valuable crop contributing to the economic transformation of the households involved in its cultivation. The importance of sunflowers arose from increased national and global demand for vegetable oils, attractive prices and use of improved seeds by farmers. In 2015, for example, the country needed 300,000 tonnes of edible oil, but only 120,000 tons (40 per cent) was sourced from within the country, the rest being imported. In order to reduce the country's dependence on imports, domestic edible oil production has had to be increased, which among other crops also requires an increase in the production of sunflower seed. Due to increased demand by small-, medium- and large-scale vegetable oil-producing enterprises, production of oil seeds recorded a remarkable increase from 418,000 tons in 2008 to over 2,800,000 tons in 2014 (URT, 2015). This increase was due to sustainable market demand, favourable weather conditions in sunflower-growing areas, the provision of subsidies for improved seeds to farmers (especially in Singida and Dodoma regions) and the training of farmers in good agricultural practices (URT, 2015). Production under contract farming was introduced in 2007/08 in the central zone in an effort to ensure sufficient supplies for the vegetable oil-producing industry. Contracts are fairly rudimentary and mostly restricted to the provision of seeds of varieties with a high oil content and agreements to purchase at current market prices.

Agricultural production and marketing arrangements in a historical perspective

This section provides a historical overview of the policy framework and strategic interventions that support agricultural development in Tanzania, divided in three distinct periods: pre-liberalisation, characterised by profound state interventions in the agricultural sector (independence to the mid-1980s); post-liberalisation (mid-1980s to the end of the 1990s), where the role of the state in crop production and marketing was significantly reduced, paving the way for private-led development; and a more recent period (2000s to the present) characterised by 'public–private partnerships' and contract farming as a new institutional tool of agricultural modernisation.

State-led intervention through marketing boards and cooperatives

A few years after the country gained independence in 1961, it adopted a state-controlled economic system in which all major economic activities, including agricultural marketing, were placed in the hands of the state. The salient features of this period were government interventions in the market through the establishment of pan-territorial and pan-seasonal producer prices for all major agricultural products, crop-marketing monopolies (including exports) granted to cooperatives and/or parastatal marketing boards and the extension of input credit to smallholders, with crop development and extension services provided by state agencies (research centres and district extension officers).

Until the mid-1970s, crop marketing was organised through a three-tier system consisting of primary cooperative societies, regional cooperative unions and parastatal crop-marketing boards. While the marketing boards were responsible for purchasing crops from the regional cooperative unions and the sale of crops on both the domestic and export markets, the cooperative unions were responsible for coordinating input supply and credit recovery, as well as the collection of crops from primary cooperative societies. By 1976 there were about 1,300 primary cooperative societies in Tanzania in almost all regions, although they were weaker in areas with a modest production of crops for export. A number of major problems led to the abolition of this three-tier system in 1976. First, the cooperatives were perceived to be dominated by large and relatively well-off farmers, who discriminated against poorer farmers, for example, in relation to input and loan allocations. This led to dissatisfaction with the cooperative unions from a majority of the smallholder farmers. Second, there was embezzlement of funds and corruption on the part of the managers in collaboration with the better-off farmers, as well as inefficiency and mismanagement of crop marketing by the cooperatives (Msambichaka *et al.*, 1982; Ponte, 2002).

As a consequence, the regional cooperative unions were formally abolished in 1976, and the primary purchasing and processing of crops came under the direct management of the Tanzanian crop authorities. These had a statutory monopoly over marketing of their respective crops and were responsible for all aspects of crop development and marketing. In conjunction with the then Tanzania Rural Development Bank, the crop authorities assisted in the provision of loans and the recovery of input credits, costs related to research and extension, working capital for crop procurement from villages, crop processing and marketing. Donor support to agricultural development during this period was substantial, with loans provided by the World Bank in particular being channelled through these crop authorities. Each of the major crops studied in this book had its own authority, namely the Tobacco Authority of Tanzania, the Tanzania Cotton Authority and the Sugar Board. In addition, the General Agriculture Products Exporting Company (GAPEX) was responsible for marketing relatively minor crops at that time, including sunflower (Ellis, 1982; Msambichaka *et al.*, 1982).

Apart from the Ministry of Agriculture, the controlling instruments for the crop authorities were (1) the board of directors, to whom the authorities were supposed to present annually audited accounts and performance reports of their operations; (2) the banks, and especially the National Bank of Commerce, the only commercial bank at that time which provided the crop authorities with working capital for which financial statements had to be shown; and finally (3) the farmers, who, despite the monopoly powers of the crop authorities, decided independently to grow the respective crops. The abolition of the regional cooperative unions and the important role of the crop authorities in coordinating crop development and marketing were perceived to allow better provision of incentives to farmers through price policy, organised input distribution and credit facilities. Besides, this was also a way for the state to secure some share of the marketing profit to run the parastatals, including facilitating them to provide services, especially extension service, to the farmers. Given the controlled foreign exchange market in Tanzania, the state control of trade and marketing ensured that foreign exchange was transferred back to the nation from the exported outputs (Coulson, 2013).

However, in 1982 regional cooperative unions and a three-tier system were reintroduced to address at least three of the challenges facing the crop authorities. First, the crop authorities' operational costs were substantial as a result of over-staffing. It was estimated, for example, that in 1980 the crop authorities were overdrawn to the tune of 5 billion Tsh, equivalent to three times the total farm-gate value of their purchases from farmers in that year (Ellis, 1982). Given that export producer prices were based on residual formulas, the increase in the operational costs of the parastatals suppressed producer prices, creating a very large margin between producer prices and export prices (Ellis, 1982). Second, even though the major cash crops in Tanzania are concentrated in different parts of the country, the involvement of the crop authorities in dealing with all aspects of crop development and marketing did not match their organisational and resource capacities. This probably explained why some farmers did not receive extension services. Third, there were conflicts between farmers and the crop authorities over the long delays in collecting crops from farmers and subsequent delays in payment. In addition, farmers had a feeling that the crop authorities were not accountable to them and therefore did not care about their interests (Verhagen, 1980). The resulting decline in smallholder crop production was quickly felt, as the government faced a shortage of foreign exchange (Maghimbi, 2010).

The reborn regional cooperative unions were placed under the then newly established Ministry of Local Government and Cooperatives, which was made responsible for the proper development of cooperative unions and district councils (Maghimbi, 2010). Some of the crop authorities, like those for tobacco, were reorganised into processing and marketing boards, although in some cases (for example, cotton) the cooperatives resumed the processing function (ginning). The functions of the cooperatives were thus more or less the same as they had been in the pre-1976 era (Chambo, 1993). The reintroduction of the cooperatives also implied that processing firms and marketing boards did not have a direct link with the producers so that extension advice was now squarely in the hands of the government, which had both fiscal and efficiency implications. First, there was a need for close coordination between the cooperatives without processing facilities and traders, on the one hand, and the processing firms, on the other hand, to ensure that firms obtained raw materials in the desired quantities and qualities. Second, the government had to increase allocations and maintain a high level of expenditure in agriculture in order to offer effective extension services (Maghimbi, 2010).

Inappropriate policies are said to have caused low growth in the agricultural sector in the late 1970s and early 1980s. These policies included the government's focus on import substitution industrialisation (ISI), supported by a controlled exchange rate and a taxation regime for the agriculture sector based on an overvalued exchange rate and export taxes. Taken together, this led to a severe decline in Tanzania's economic performance during the early 1980s. Exports lagged far behind the imports that were required to support the ISI strategy, resulting in unsustainable balance of payment deficits. Internally, revenue collection fell short of expenditure partly because of government support to the loss-making parastatals and partly because some of the socialist policies of the time stressed the provision of free services, including education and health. The structural imbalance and declining agricultural growth rates forced Tanzania to start a reform process leading to it abandoning the state-controlled regime and adopting a market economy. This came into effect from the 1984/85 financial year (World Bank, 1994).

Liberalisation and structural adjustment programmes

Agriculture was at the core of the 'Economic Recovery Programme' that was implemented from 1986. Agricultural reforms were expected to have a positive impact on the whole economy as the agricultural sector accounted for the majority of total export earnings and provided a livelihood for the majority of the workforce in the country. Based on the core recommendations of the World Bank, the 'Agricultural Policy of Tanzania' was formulated in 1983 (Ponte, 2002). The Economic Recovery Programme and subsequent Structural Adjustment Programmes (SAPs) aimed to reduce the fiscal deficit while also stimulating growth in the agricultural sector by eliminating price distortions and bringing in an efficient market system. The latter were expected to improve production responses, promote agricultural growth, increase the incomes of poor rural households and contribute to poverty reduction. The roles of the parastatals and cooperative unions in crop marketing were significantly reduced or redefined. The crop parastatals retained their regulatory functions only, including issuing trading and processing licences to private companies and ensuring competition, while the regional cooperative unions were reconstituted on criteria economic viability, meaning that they could (re-)register under the new cooperative act provided they were solvent (Cooksey, 2011).

Most domestic market controls on food crops were abolished in 1989, including restrictions on the marketing and transportation of food grains. The National Milling Corporation maintained its obligations as the buyer of last resort and the manager of food security stocks, but private traders rapidly replaced the parastatal crop authority in food marketing (Ponte, 2002). Producer prices for the main agricultural products were liberalised between 1991/92 and 1993/94. In addition, subsidies for agricultural inputs such as fertiliser, introduced in the 1970s and early 1980s, were phased out (World Bank 1994). The liberalisation of the export crop sector was initiated in the early 1990s. The agricultural policy reforms pertaining to export crops were spelled out in the Economic Recovery Programme II – Economic and Social Action Programme (1989–92). These reforms included the removal of the fixed producer price system (pan-territorial and pan-seasonal prices) and the privatisation of the former parastatal processing companies (e.g. sugarcane- and tobacco-processing companies). New statutory crop regulations specified that private companies were allowed to enter the market at every stage, from the primary purchase to the export of crops, provided that licences were obtained from the public authorities.

The dismantling of the former parastatal marketing systems and liberalisation of the crop markets thus paved the way for the participation of cooperative unions, private companies and traders in the processing and marketing of all agricultural crops. However, several cooperative unions were liquidated as part of the cooperative reforms, and the role of cooperative unions in crop-processing and marketing significantly declined further during the 1990s, as they had to operate in a competitive environment and thus behave as real business entities operating on commercial lines along with private traders and/or processers.

A number of agricultural policies and strategies were developed to support the SAPs. The Agricultural and Livestock Policy of 1997 provided a policy framework for the reforms of the agricultural sector. The salient features of the Agricultural and Livestock Policy were the removal of state monopolies in the export and import of agricultural goods and the increased role of the private sector as an engine of growth in crop production, processing and marketing. The policy also aimed to improve livelihoods in rural areas through increased income generation from agriculture and livestock production, processing and marketing (URT, 1997, 1999).

Generally, implementation of SAPs and liberalisation of the agricultural sector were aimed at providing competition and promoting efficiency by, among other things, allowing markets to determine crop prices. The assumption underlying the market reforms was that competition between companies would lead to improved quality of service, for example, more timely provision of inputs, faster payments for crop purchases and better prices to producers. However, the results of SAPs on agricultural production responses and increased incomes for rural households are contested and vary tremendously depending on the specific crop in question. Market reforms and the introduction of private traders and processors have usually resulted in more efficient flows within the marketing system, with prompter cash payments and a higher share of the export price being paid to farmers (Friis-Hansen, 2000; Ponte, 2002). However, while production of tobacco (and cashew nuts) has increased significantly since liberalisation, most other crops seemed to benefit from a rise in production for around two seasons after the agricultural reforms, followed by stagnation in the following seasons in the 1990s. According to Ponte (2002), estimates of crop production figures and agricultural growth seemed to be similar before and after the implementation of reforms.

Although it is difficult to draw firm conclusions concerning the effects of SAPs on agricultural performance in Tanzania, low rates of land and labour productivity remained one of the critical challenges in the agricultural sector in the later part of the liberalisation period. This is said to have been caused by multiple factors, but mainly inadequate finance to obtain productivity-enhancing inputs and appropriate technology (e.g. improved seeds), as well as the limited availability of support services (research and extension) (URT, 2011a). Prior to liberalisation, input credits were provided to farmers against future sales to the parastatal marketing boards, often through cooperative unions and/or primary cooperative societies. After liberalisation, it was expected that cooperative unions and private companies would procure chemicals (fertiliser, pesticides, insecticides) on their own account for redistribution to smallholders on credit or cash terms. However, as fertiliser subsidies were gradually removed in the 1990s, prices rose, while consumption declined significantly (Ponte, 2002).

This was caused by the fact that, following the reforms, the cooperative unions were increasingly unable to obtain inputs on credit themselves. Given the increased competition and increasing scope for side-marketing by farmers, the cooperative unions therefore stopped credit provisions to smallholders. In addition, market failure in input distribution has been one of the salient characteristics of liberalisation, notably in remote areas with poor transport infrastructure. Most private companies and traders have been reluctant to procure and supply inputs to farmers because of low levels of liquidity in combination with a lack of access to credit at commercial banks, which view agricultural lending as high-risk lending with low returns (URT, 2010a).

The post-SAP period saw a rapid decline in donor support for agriculture. The government's expenditure on agricultural development declined significantly in the 1990s as a consequence of fiscal savings. At the end of the 1990s, agriculture accounted for only 3.5 per cent of the state budget (Cooksey, 2012).

Liberalisation and institutional intervention

The more recent policy formulations and initiatives aimed at agricultural transformation in Tanzania reflect the so-called post-Washington consensus that emerged in the aftermath of the SAPs in the international donor community. Broadly speaking, the post-Washington consensus takes liberalisation and financial stringency as given, but also accepts that agricultural development requires a broader series of interventions (Friis-Hansen, 2000). During the 2000s, the Ministry of Agriculture, Food Security and Cooperatives was given a more important role in promoting market development and facilitating the emergence of specific institutions to support agricultural growth, while the private sector was still expected to perform key roles in agricultural transformation.

The restoration of the government's priority to allocate public investments in support of agriculture originated in the renewed political (and donor) interest in agriculture as a driver of broad-based poverty reduction and growth. At the regional level, this is reflected in the African Union's strategic framework for socio-economic development, the New Partnership for Africa's Development (NEPAD), and its key programme component, the Comprehensive Africa Agricultural Development Programme (CAADP), launched in 2003. Through this, African governments committed themselves to increasing national budgetary allocations to agriculture (10 per cent in 2015), ensuring a 6 per cent annual increase in agricultural GDP and promoting the agribusiness sector through the CAADP (URT, 2011c). Public spending on agriculture in Tanzania increased in absolute and relative terms in the 2000s, though it has remained below the target of 10 per cent of public spending as agreed in the CAADP (Cooksey, 2012: 7).

The government has developed various policies and strategies to stimulate agricultural production as a means to reduce poverty and spur national economic growth, including the Agricultural Sector Development Strategy 2001, the National Strategy for Growth and Reduction of Poverty (I and II), the Agricultural Sector Development Programme, the so-called 'Kilimo Kwanza', the Agricultural Marketing Policy, the National Agricultural Policy and the National Five Year Development Plans. They are reviewed in the remaining part of this section.

The Agricultural Sector Development Strategy 2001 was developed as an integral part of the reforms to the agricultural sector with the purpose of addressing some of the challenges in transforming the sector. Among the focal interventions were the commercialisation of agriculture through increasing access to inputs in rural areas (as a way of contributing to productivity increases) and promotion of agro-processing and rural industrialisation. The latter is linked to the promotion of farmers' organisations for empowering farmers and forging partnerships between smallholders and agribusinesses through contract farming or other arrangements (URT, 2001). The aim was to provide access to markets, inputs and technology to smallholder farmers, as well as to ensure a reliable supply of crop produce to the processors.

The National Strategy for Growth and Reduction of Poverty (NSGRP) – popularly known as MKUKUTA – was implemented as part of the country's debt relief granted by the IMF (URT, 2005b, 2010c). NSGRP I and II (2005–10 and 2011–15) take their point of departure in the National Development Vision of 2025. They focus, among other things, on industry-specific targets for agricultural production and emphasise the need for increased agricultural productivity and profitability, as well as the secure and facilitated marketing of agricultural products. The state is envisaged as being responsible for the management of food security stocks in the country, as well as for strengthening the institutional framework that manages agricultural development. The private sector is envisaged as performing two key roles: to invest in the agricultural sector (including value-addition activities), and to provide agricultural support services (including research, extension and inputs services).

While the NSGRP (I and II) and the National Development Vision of 2025 are mainly strategic papers, the Agricultural Sector Development Programme (ASDP) and 'Kilimo Kwanza' have had a more direct and tangible impact on the development of the agricultural sector. The ASDP was formulated and adopted in 2006 by the Ministry of Agriculture, Food Security and Cooperatives and became the basis of the government's budgetary and general resource allocations in the agricultural sector in 2006–13. The stated purpose of the ASDP is to achieve an annual agricultural growth rate of 5 per cent. This is to be achieved by state-led investments in agricultural support institutions that aim to improve the marketing of inputs and outputs and provide an 'enabling environment' to encourage private market-oriented investments. More specifically, ASDP aims to:

1 Give farmers better access to and use of agricultural knowledge, technologies, marketing systems and infrastructure (including irrigation schemes and water management).
2 Increase private-sector investment in agriculture based on an improved regulatory and policy environment.

(Cooksey, 2012: 9; URT, 2005a)

The ASDP was 'basket-funded' by major international development agencies (URT, 2005a; Cooksey, 2012). It included two components and associated grants. First, there was a local component with grants to local governments for investments in infrastructure and agricultural services (notably extension), capacity-building and the empowerment of farmers' organisations, local government and the private sector. It was planned to have this component implemented by local authorities through district agricultural development plans. The revival of farmers' organisations or cooperatives as an institutional device for agricultural transformation was to be promoted by district government authorities. Farmers' organisations were considered important platforms for the more efficient supply of inputs and the dissemination of extension services and savings and credit groups. Second, there was a national component aimed at financing agricultural research and extension service activities, national-level infrastructure, market development activities and policy development and planning (Cooksey, 2012).

The ASDP and subsequent policy strategies to increase commercialisation and enhance productivity in the agricultural sector initiated a process of change with regard to the levels and forms of state intervention in the sector and the role of the private sector in fostering agricultural transformation. This is reflected in the so-called 'Kilimo Kwanza' (Agriculture First), launched by the Tanzania

National Business Council in 2009 as a public–private sector plan for agricultural transformation (TNBC, 2009). The 'Kilimo Kwanza' resolution is a market-led vision for the transformation of the agricultural sector driven by private-sector participation and investments and supported by the state. The state is envisaged as allocating more than 10 per cent of the national budget to agriculture to finance the implementation of the strategic framework (TNBC, 2009). The private sector is encouraged to invest in land (including large-scale farming agriculture) and agriculture-related infrastructure and service delivery. The framework consists of ten pillars concerned with mechanisms for improvements to the institutional and fiscal environment to encourage and support private-sector investments in the agricultural sector. Contract farming is endorsed as one of the ten pillars for agricultural development, and the resolution also advocated 'fast-track' legislation to support contractual arrangements in the agricultural sector.

The Agricultural Marketing Policy of 2010 aims to put in place an orderly marketing system which guarantees producers reliable markets for their crops, especially in remote rural areas, and taking into account perishability and other risks associated with crop cultivation. The strategy provides for a proactive role by the government, particularly in the event of output market failures (URT, 2010b). In addition, crop-specific laws have been amended to increase the role of the parastatal boards in regulating crop development and making it possible for the boards to enter the market as commercial actors financially supported by the state (Cooksey, 2011: 68). Hence, the state, through the crop-specific marketing boards, has regained some of its pre-liberalisation 'functions' in regulating and promoting agricultural development. However, the recent 'reversal' to state-led agricultural interventions and investments emphasises the role of the private sector as the key driver of agricultural transformation. For example, the Agricultural Marketing Policy of 2010 includes instruments to encourage the development of the private sector by creating a conducive environment for private-sector participation in agricultural production, processing and marketing. In addition, the National Agricultural Policy of 2013 provides instruments to strengthen the public agricultural support services, including extension and training and support to inter-sectoral coordination and linkages, to increase the efficiency and effectiveness of the sector.

Contract farming and the creation of farmers' organisations are explicitly addressed in the recent National Five Year Development Plans as two core means to improve the availability and utilisation of modern agricultural inputs, mechanisation and productivity in the agricultural sector. The most recent Five-Year Plan, covering the period from 2016 to 2021, focuses on increased industrialisation, but emphasises the agricultural sector as the main pillar for economic growth. The strategy and planned interventions aim to increase the growth rate to 7.6 per cent in the agricultural sector by 2020 (URT, 2015, 2016a). The strategic interventions include the increased use of modern technologies and extension services in order to increase productivity, the lengthening and deepening of value chains, improvements to infrastructure (transportation, energy, water supply and irrigation, and warehousing, including cold chains) and better access to financial services. Finally, it stresses deliberate efforts to develop contract farming or out-grower schemes and the promotion of farmers' groups to complement the raw material sources of manufacturing industries.

Conclusion

Despite its declining share of the country's gross domestic product, agriculture continues to be a significant sector in Tanzania as a source of livelihood for a significant proportion of the population. Cotton, sugarcane, sunflower and tobacco are all given priority in the national agricultural development initiatives because of their contribution to both national and local economic development and industrialisation processes.

A plethora of policies and strategy frameworks have been developed in an attempt to spur the contribution of the sector to the wider economy, with more outward-looking policies that encourage the private sector to play a leading role in production and marketing activities. The more recent policy initiatives and interventions in particular emphasise private-sector investments in agriculture based on an improved regulatory environment. Hence, it is notable that earlier state regulatory mechanisms have experienced a partial revival, albeit in a new context of economic dynamics conditioned by private capital.

The agricultural sector has faced a number of chronic challenges, leading to repeated strategic interventions in the sector by the state. Hence, the policies and strategic frameworks have continually stressed the need to maximise agricultural output and productivity. Low productivity of land and labour remain one of the critical challenges, and there are calls for improvements in research and extension services, as well as in access to and the appropriate use of improved seeds, fertilisers, chemicals and post-harvest technologies. However, limited access to financial services by smallholder farmers constrains their ability to purchase farm inputs and tillage equipment. Commercial banks and other financial institutions consider loans to the smallholder side of the agricultural sector to entail high risks with low returns, not least because of the costs associated with the management of loans to a multitude of smallholders scattered over large areas with little or no collateral. While the number of producers' organisations (such as cooperative societies, saving and credit organisations and farmers' organisations) has increased, their role in facilitating access to financial services is generally very marginal. In addition, agricultural marketing has been constrained by several factors, including an inadequate institutional, legal and regulatory framework. Limited agro-processing industries in rural areas have discouraged smallholder production and resulted in high post-harvest losses.

Contract farming as a concept and institutional device for agricultural transformation was seriously endorsed in the 'Kilimo Kwanza' resolution, supposedly backed by specific legislation to support contract farming arrangements in the agricultural sector. Contract farming and the creation of farmers' organisations are also explicitly addressed in the National Five Year Development Plans, the most recent one stressing the agricultural sector as the main pillar of economic growth. The plans strongly advocate the development of out-grower schemes (contract farming) to complement the sources of raw materials for the appropriate industries. Hence, there is currently strong policy support for contract farming arrangements in areas and crops where this can work because of the potential this has for addressing some of the chronic challenges that have hindered the transformation of the agricultural sector. This justifies further research that aims to increase our understanding of the potentials and limitations of contract farming.

References

Bank of Tanzania (2016). *Monthly Economic Review, May 2016*. Online: www.bot.go.tz/Publications/MonthlyEconomicReview/MER May 2016.pdf. Accessed 19 September 2016.

Chambo, S. A. (1993). The Impact of Farmers on Government Policy in Cooperative in Tanzania. Paper presented at the workshop 'How to Increase the Level of Communication Experiences from Farmers, Professional and Policy Makers'. Sokoine University of Agriculture, 6–9 September 1993.

Coulson, A. (2013). *Tanzania: A Political Economy* (2nd edn). Oxford: Oxford University Press.

Cooksey, B. (2011). Market Reform? The Rise and Fall of Agricultural Liberalisation in Tanzania. *Development Policy Review*, 29, 557–581.

Cooksey, B. (2012). *Politics, Patronage and Projects: The Political Economy of Agricultural Policy in Tanzania*. FAC Political Economy of Agricultural Policy in Africa, Working Paper 040. Online: www.future-agricultures.org. Accessed 5 December 2016.

Ellis, F. (1982). Agricultural Price Policy in Tanzania. *World Development*, 10(4), 263–283.

FAO (2013). *Review of Food and Agricultural Policies in the United Republic of Tanzania 2010–2011: Country Report*. Rome: Food and Agricultural Organisation.

Friis-Hansen, E. (2000). *Agricultural Policy in Africa after Adjustment*. Copenhagen: Centre for Development Research.

Maghimbi, S. (2010). *Cooperatives in Tanzania Mainland: Revival and Growth*. COOP Africa Working Paper No. 14. Rome: International Labour Organization.

Msambichaka, L. A., Ndulu, B. J. and Amani, H. K. R. (1982). *Agricultural Development in Tanzania: Policy, Evolution, Performance and Evaluation – The First Two Decades of Independence*. Bonn: Frierich-Abert-Atiftung.

Ponte, S. (2002). *Farmers and Markets in Tanzania*. Oxford: James Currey.

TNBC (2009). *Kilimo Kwanza: Towards Tanzania Green Revolution*. Dar es Salaam: Tanzania National Business Council.

UNDP/URT (2015). *Tanzania Human Development Report 2014: Transformation for Human Development*. Economic and Social Research Foundation. Dar es Salaam: Economic and Social Research Foundation, United Nations Development Programme and Ministry of Finance.

URT (1997). *Agricultural and Livestock Policy*. Dar es Salaam: Ministry of Agriculture and Cooperative Development, United Republic of Tanzania.

URT (1999). *Tanzania National Development Vision*. Dar es Salaam: President's Office, Planning Commission, United Republic of Tanzania.

URT (2001). *The Agricultural Sector Development Strategy, 2001*. Dar es Salaam: Ministry of Agriculture, United Republic of Tanzania. Online: www.tzonline.org/pdf/agriculturalsectordevelopmentstrategy.pdf. Accessed 19 September 2016.

URT (2005a). *Agricultural Sector Development Programme, 2005*. Dar es Salaam: Ministry of Agriculture, United Republic of Tanzania. Online: www.kilimo.go.tz/.../ASDP/FINAL/2025/2005. Accessed 21 September 2016.

URT (2005b). *National Strategy for Growth and Reduction of Poverty*. Dar es Salaam: Vice President's Office, United Republic of Tanzania.

URT (2010a). *Agricultural Sector Review and Public Expenditure Review 2008/2009*. Ministry of Agriculture, United Republic of Tanzania. Online: www.agriculture.go.tz/publications/.../ASR-PER/REVIEW. Accessed 19 September 2016.

URT (2010b). *Agriculture Marketing Policy*. Dar es Salaam: Ministry of Industry, Trade and Marketing, United Republic of Tanzania.

URT (2010c). *National Strategy for Growth and Reduction of Poverty Agricultural Sector Development Programme II*. Dar es Salaam: Ministry of Finance and Economic Affairs, United Republic of Tanzania.

URT (2011a). *ASDP Performance Report 2009/10*. Dar es Salaam: Ministry of Agriculture, Food Security and Co-operatives, United Republic of Tanzania. www.agriculture.go.tz/.../ASDP/.../ASDP/Performance. Accessed 21 September 2016.

URT (2011b). *National Five Year Development Plan 2010/11–2015/16*. Dar es Salaam: President's Office, Planning Commission, United Republic of Tanzania.

URT (2011c). *Agricultural Sector Review/Public Expenditure Review 2010/2011*. Dar es Salaam: Ministry of Agriculture, Food and Cooperatives, United Republic of Tanzania.

URT (2014). Speech by the Minister of State, Social Relations and Coordination, Honourable Stephen Masato Wasira Presenting to the National Assembly The National Economic Survey for 2013 and Annual Development Plan for 2014/15. 12 June 2014, United Republic of Tanzania.

URT (2015). Speech of the Minister of Agriculture, Food and Cooperative Christopher K. Chiza on the Budget for the Ministry of Agriculture, Food and Cooperatives for the Year 2015/16. United Republic of Tanzania.

URT (2016a). *1914/15 Annual Agriculture Sample Survey: Dar-es-Salaam and Zanzibar*. Ministry of Agriculture, Livestock and Fisheries; Ministry of Agriculture and Natural Resource, Zanzibar; Ministry of Livestock and Fisheries, Zanzibar; President's Office, Regional Administration and Local Governments; Ministry of Industries, Trade and Investment; National Bureau of Statistics and the Office of the Chief Government Statistician, Zanzibar. United Republic of Tanzania.

URT (2016b). *National Five Year Development Plan 2016/17–2020/21*. Dar es Salaam: Ministry of Finance and Planning, United Republic of Tanzania.

Verhagen, K. (1980). Tanzania Rural Development Policy 1975–78. *Development and Change*, 2, 285–295.

WHO (2003). *World Health Organization Framework Convention on Tobacco Control*. Geneva: World Health Organization.

World Bank (1994). *Tanzania Agriculture: A World Bank Country Study*. Washington, DC: World Bank.

Part II

Contract farming and value chain dynamics

4 Evolving governance structures and contract farming in the tobacco value chain in Tanzania

Bahati M. Ilembo, Joseph Andrew Kuzilwa and Marianne Nylandsted Larsen

Introduction

The global tobacco industry is dominated by a handful of large international cigarette manufacturers that have amassed significant financial, political and social influence in the global value chain (GVC) for tobacco. These branded cigarette manufacturers are what Gereffi (1994) has referred to as 'lead firms' in the GVC in the sense that they are able to influence the activities of other actors in the GVC. All the international cigarette manufacturers have numerous strong brands in consumer markets based on unique mixtures of tobacco blends, tightly coordinating their supply chains to ensure the origin of the tobacco blends and the conditions under which they have been produced. Performance requirements are transmitted to tobacco-processing companies and merchants in tobacco-growing countries, including conformity with good agricultural and labour practices (for elaboration on the latter, see Chapter 13, this volume). In order to secure consistency and a reliable supply base, contract farming arrangements between tobacco-processing companies and tobacco-growers often forms part of the tightly coordinated value chain.

Over 40 countries produce tobacco, but the majority of the world's flue-cured tobacco originates from around nine major tobacco-producing countries (Brazil, China, United States, Malawi, Mozambique, Tanzania, Zimbabwe, Italy and Turkey), and over 80 per cent of the world's tobacco production is used for cigarettes (FAOSTAT, 2014; Agricultural Marketing Resource Centre, 2012). Global tobacco trade flows are determined by trends in demand and supply, but they are also influenced by trade policies such as tariffs and non-tariff barriers, export promotion and international and national regulations seeking to address the health risks associated with smoking. Notably in recent decades, the branded cigarette manufacturers and the tobacco industry as a whole have been confronted with awareness and other campaigns against cigarette consumption by both public and private anti-smoking groups (Joossens and Gilmore, 2013).

These health concerns have spurred various regulatory measures: globally the World Health Organization (WHO) has worked extensively to produce uniform rules that seek to protect present and future generations from the devastating social, environmental and economic consequences of tobacco consumption and exposure to tobacco smoke (WHO, 2003). In 2003, the WHO launched a protocol for member countries to adopt strong tobacco control policies through the provision of rules that govern the production, sale, distribution, advertisement and taxation of tobacco. Nationally, too, most countries have introduced a regulatory framework for tobacco control with the general objective of reducing tobacco use and its consequent harm in various ways. This often also includes publicly available information about the risks of using tobacco products and of exposure to second-hand tobacco smoke, as well as about the benefits of quitting smoking. For example, these kinds of regulations and publicity strategies against smoking are entailed in Tanzania's Tobacco Products Act (2003), which also prohibits the advertising of tobacco products in the media and sales of tobacco products to people under the age of 18. In addition, smoking is prohibited inside public transport, schools, hospitals and other public places (see also Li and Guindon, 2013; Warner and Mackay, 2006). Hence, the cigarette manufacturers are subjected to international and national rules and regulations which they need to address in their business operations, supply-chain coordination and marketing strategies. For example, with regard to the latter, the tobacco industry has been forced to introduce health warning statements on cigarette packs in many countries and regions, including the United States and European Union.

Despite global and national efforts to eliminate or at least reduce tobacco consumption due to its associated health risks, not to mention its negative environmental and social impacts in tobacco-growing countries, most African governments have promoted smallholder tobacco-farming as a way to alleviate poverty, especially in areas where there are no alternative high-value crops. Tobacco production in Tanzania has been referred to as a success story because it now involves a substantial number of smallholder farmers in export-oriented crop cultivation based on contract farming. Contract farming appeared first in the early 1960s as part of a large state-sponsored rural development project focusing on the promotion of smallholder-based export crops (Boesen and Mohele, 1979). Although accurate data are difficult to obtain, it is estimated that more than 100,000 smallholders are currently involved in the production of tobacco in Tanzania, with about 60 per cent located in Tabora Region in the west of the country. According to the Bank of Tanzania, the crop has ranked first in foreign exchange earnings among the major commodities for the last three years (Bank of Tanzania, 2016). Hence, the crop is important in sustaining the livelihoods of numerous people working along the value chain, from green-leaf cultivation to processing dried leaf into tobacco blends. In addition, tobacco production contributes to national economic development through vital foreign-exchange earnings (URT, 2013b, 2016).[1]

This chapter explores the (changing) relationships between the two main groups of actors – smallholders and tobacco-processing companies – in the Tanzanian part of the global value chain for tobacco. The chapter begins with an elaboration of the notion of governance provided by the GVC literature

(Gereffi, 1994; Gereffi *et al.*, 2005). This is followed by an analysis of the contractual system and the relationships between smallholders and tobacco-processing companies in three distinct periods of tobacco contract farming in Tanzania:

1 Pre-liberalisation (1960–95), when a parastatal coordinated the entire tobacco value chain in Tanzania and the contract farming arrangements between smallholders and parastatal tobacco-processing companies.
2 Post-liberalisation (1995–2008), when the contract farming scheme was taken over by two private tobacco-processing companies, both subsidiaries of major international tobacco-leaf merchants.
3 The recent era (2009–14), when the private tobacco-processing companies were restricted from involvement in the input credit system, and the regional cooperative unions, representing all primary cooperative societies in a designated geographical area, assumed the role of coordinating input credit supply to smallholders.

This chapter argues that the changing governance structures in the Tanzanian tobacco value chain are not primarily a result of chain-internal dynamics, such as changes in supplier capacities or the enhanced ability to codify information about product or process requirements. Lead firms' strategies and state institutions are intertwined and shape the tobacco value chain's governance structures in Tanzania. In addition, the chapter discusses the impacts of the more recent state-imposed changes to contractual arrangements on the sustainability of primary cooperative societies as the prime entry point for coordinating contract farming between smallholders and private tobacco-processing companies. The chapter ends with a conclusion.

The chapter is based on findings from a survey of 300 smallholders conducted in Urambo District in 2012. In addition, in-depth interviews were carried out with two tobacco-processing companies located in Morogoro, the association of Tanzanian tobacco traders, the regional cooperative union (Tabora region) and ten primary cooperative societies, as well as smallholder farmers within each interviewed primary cooperative society in Urambo District in 2013. In addition, interviews were carried out with government staff at various levels (for a description of fieldwork methods, see Ilembo, 2015).

Governance in global value chains

A global value chain can be seen as a 'set of inter-organisational networks clustered around one commodity or product, in which networks are situationally specific, socially constructed and locally integrated' (Gereffi, 1994: 2). Gereffi identifies four analytical dimensions of GVCs: (1) the input–output structure or configuration, that is, a set of economic activities linked together in a sequence of value-adding economic activities; (2) the geographical coverage of value chains: the spatial dispersion or concentration of the production and distribution network; (3) the chain governance structure, that is, power relations that influence how financial, material and human resources are allocated and flow within the chain (Gereffi, 1994: 97); and (4) the institutional framework through which national and international conditions and policies shape or influence processes and relationships between actors at different segments in the value chain (Gereffi, 1999).

Of these four dimensions, the notion of a governance structure has received most attention, since this is where the key notions of entry barriers and chain coordination appear in the analytical framework and where the distinction between 'producer-driven' and 'buyer-driven' value chain governance is introduced. Governance draws attention to the vertical coordination of economic activities through non-market relations (i.e. inter-firm linkages) and the nature and scope of these relationships between actors in different segments of the value chain. Gereffi (1994: 97) defines governance as 'authority and power relations that determine how financial, material and human resources are allocated and flow within a chain'. So-called lead firms have the power to determine or influence the division of labour between actors in the chain through, for example, the ability to out-source (or externalise) lower value-added functions up and/or downstream in the chain while retaining or incorporating those with higher added value. In addition, lead firms define performance requirements for their suppliers, including product details, availability and delivery accuracy, procedures for compliance with standards, and quality control management. The power of lead firms stems from two attributes, namely a strong market presence, and their position in the segment of the chain that has the highest barriers to entry. Governance structure, as opposed to arm's-length relations, arises from the need to coordinate internationally dispersed activities and, at the same time, address suppliers' performances according to the requirements set by the lead firm (Gereffi, 1999; Humphrey and Schmitz, 2002).

Two main types of governance were originally proposed by Gereffi (1994): 'buyer-driven' and 'producer-driven' value chains. Producer-driven chains are said to characterise capital- and technology-intensive industries (e.g. industries like automobiles and aircraft), where large industrial enterprises or transnational companies play the central role in controlling the production system. Buyer-driven chains differ from producer-driven chains in that they typically have lower barriers to entry into production. These chains are usually found in more labour-intensive industries (e.g. garments and footwear), where production functions are entirely outsourced and in which lead firms – large retailers, branded manufacturers, branded merchandisers – play the pivotal role in setting up decentralised production networks in a variety of exporting countries.

Gereffi's conceptualisation of two ideal forms of governance has been criticised on several counts. For example, research on the electronics industry (Sturgeon, 2002) and on the cocoa–chocolate complex (Fold, 2002) suggest that, although these chains are highly buyer-driven, different types of lead firms are important for how the chains are governed. In addition, critics have opposed the empirically derived forms of the two archetypical governance structures and the heavy bias towards buyer-driven chains in the GVC literature (Gibbon and

Ponte, 2005). Such debates have led to efforts to refine definitions of governance in GVCs in terms of how lead firms set, measure and enforce the parameters under which others in the chain operate (Humphrey and Schmitz, 2002; Gibbon and Ponte, 2005: 80–81). Gereffi *et al.* (2005) have therefore proposed a theoretically grounded set of more diversified types of governance inspired by transaction-cost theory. The prime focus is on the relationship between lead firms and first-tier suppliers. The authors identify three parameters for determining the actual form of governance: (1) the complexity of inter-firm transactions of knowledge and information. More complex transactions require greater interaction among actors in the GVC and hence stronger forms of governance than simple price-based transactions; (2) the codifiability of transactions, which refers to the ability to codify knowledge and information such that complex information can be shared between lead firms and suppliers; and (3) the capabilities of the suppliers in relation to the requirements of the transaction. These parameters are measured as high or low, resulting in eight combinations, three of which are ruled out in practice as meaningless (Gereffi *et al.*, 2005). The remaining five types of buyer–seller relations are claimed to constitute a specific form of governance in particular GVCs. In addition to the end-points along the continuum – market and hierarchy – three different types of governance are identified:

- Modular: characterised by high information complexity, but also high codifiability of information. The information and knowledge necessary for the transaction is somehow easy to codify, for example, through technical or product standards. Suppliers have high capabilities, and the technology used is sufficiently generic to allow suppliers to have a broad customer base.
- Relational: characterised by a low ability to codify information and high supplier capabilities. The knowledge required is often tacit, and complex information is exchanged with little codification of the product specifications. The relationship between buyer and seller is more egalitarian, and exchange of information is based on a certain degree of mutual dependence, often regulated through reputation, social ties or spatial proximity.
- Captive: characterised by high information complexity, a low ability to codify complex product specifications and low supplier capabilities. Suppliers are dependent on lead firms, which monitor but also assist their suppliers with, for example, technical or managerial assistance to enable them to perform certain functions in the production segment of the value chain.

(Gereffi *et al.*, 2005)

In this framework, the degree of explicit coordination and power asymmetry among actors increases as GVCs move from market to hierarchy. According to Gereffi *et al.* (2005), these forms of governance are not static and can evolve over time and space in a particular industry as a result of changes in one of the three key parameters. For example, changes in supplier capabilities might cause a change in governance from captive to relational. In addition, development in technology and innovation can change the abilities to codify complex information and as such influence the degree of explicit coordination required in the transactions.

Moving beyond the classification of the forms of governance that prevail in GVCs, it is important to note that the fourth analytical dimension, the institutional framework surrounding the chain, has received less attention in most of the GVC literature.[2] One notable exception is Neilson and Pritchard (2009), who focus on upgrading in the coffee and tea sectors in South India. The authors understand governance arrangements and institutional formations as being co-produced by internal, pace-bound actors and external, lead-firm actors. As such, the institutional framework should not be seen as something 'external' to the value chain, nor should chain governance be seen as 'an external structure imposing its will upon passive, spatially bound, economic actors' (Neilson and Pritchard, 2009: 43–54).

The authors analyse the GVCs for tea and coffee originating in the South Indian agricultural sector. Both GVCs are dominated by a handful of large international branded manufacturers with the capacity to influence the involvement of various economic actors along the value chains and hence reshape socio-economic outcomes at the production sites of the chains. Yet, tea and coffee estates, smallholders and workers in South India are embedded in various institutional arrangements that influence the terms of engagement within the specific GVCs. The authors stress that the restructuring processes of the GVCs for coffee and tea do not occur within a 'passive political landscape' (Neilson and Pritchard, 2009: 211). Institutional 'struggles' by actors in the South Indian tea and coffee sectors influence the capacity of producers upstream in the chain to participate in GVCs. For example, in the case of tea, the Tea Board of India and its local agency in South India has played a pivotal role in supporting smallholders' participation in the GVC. Based on their empirical analysis of the tea and coffee GVC, Neilson and Pritchard (2009) argue that institutional environments interact with chain governance structures (i.e. linkages between actors in the GVC). The evolution of specific forms of governance is a result or culmination of the institutional environments within which economic actors are embedded. Therefore, the authors suggest that chain governance structures are transformed and shaped by space, place and territory and the power of 'place-based institutions to negotiate global *governance* structures' (Neilson and Pritchard, 2009: 63–64, emphasis in original).

Contract farming in the Tanzanian tobacco value chain

State-governed contract farming (1960–95)

Smallholder tobacco production was promoted in the mid-1950s by the then Tanganyika Agricultural Corporation and the East African Tobacco Company, a subsidiary of one of the largest international cigarette companies. The

Tanganyika Agricultural Corporation was a colonial parastatal responsible for settlement schemes. Among its activities were the establishment and administration of the tobacco settlement scheme in Urambo (Boesen and Mohele, 1979: 135). The scheme was highly successful in rapidly expanding production, and an increasing number of smallholders became involved in the cultivation of tobacco leaves. After independence, smallholder tobacco production was intensified through various state initiatives. New cooperatives were established to take care of the distribution of inputs on credit terms to smallholders and the primary marketing of tobacco, while the Tanganyika Tobacco Board became responsible for purchasing, manufacturing and trading the tobacco leaves collected from the cooperatives. In the early 1970s, smallholder involvement was further strengthened by the development of so-called tobacco complexes set up by the Ministry of Agriculture and financially supported by World Bank to expand tobacco production in the country (Boesen and Mohele, 1979: 31). A new parastatal, the Tobacco Authority of Tanzania (TAT), was established to maintain all the responsibilities for the development of tobacco production, and the cooperatives were subsequently abolished in 1976.

The tobacco complexes consisted of a number of newly established villages located in close proximity, and each village acted as a complete production unit. Village (cooperative) leadership became responsible for the distribution of inputs and the collection of tobacco leaves from smallholder tobacco-growers, as well as administration of the credit and payment system. Smallholders received the necessary fungicide and blended fertiliser inputs on credit terms, the costs of inputs being deducted when smallholders sold their tobacco to the village cooperative. The tobacco delivered from the 'tobacco complex villages' was processed by the parastatal TAT at a newly built tobacco-processing plant in Morogoro. As part of the implementation of 'Ujamaa' and 'villagization', the tobacco complexes were 'filled up' with new settlers (Boesen and Mohele, 1979: 71; see also Ponte, 2002, for an account of Ujamaa and villagisation).

Taken together, the parastatal TAT coordinated the entire tobacco value chain and the contract farming arrangement between smallholders and parastatal tobacco-processing companies: it was responsible for buying tobacco leaves from smallholders through the village cooperative, keeping village records on inputs and sales, transport, storage, grading and processing leaves, as well as sales of tobacco blends.

Private tobacco-processing companies coordinating contract farming (1995–2008)

From 1995, and with the implementation of structural adjustment programmes, the state gradually started to withdraw from direct involvement in the tobacco value chain. TAT's statutory monopoly in the purchasing, processing, marketing and export of tobacco was removed, and the former parastatal processing plants were privatised in 1997. The state, through the Tanzania Tobacco Board, retained its regulatory role. These reforms paved the way for private-sector involvement in the tobacco value chain. One of the former parastatal processing plants (located in Morogoro) was sold to one of the largest international tobacco leaf companies, which subsequently modernised the plant, while another multinational company entered the market and constructed a new factory, also located in Morogoro (Mitchell and Baffes, 2002).

Under the liberalised market, tobacco production continued to be practised under contract farming arrangements between the smallholders and the tobacco-processing companies. Under these arrangements, the two tobacco-processing companies were obliged to provide inputs to smallholders on credit through primary cooperative societies (PCSs) on the condition that the growers sold all their tobacco produce to the respective PCSs. The required inputs were based on estimated needs according to land allocated to tobacco cultivation. The input loan repayment was made during the tobacco marketing. The PCSs, legally established in tobacco production areas, are the main local intermediaries between smallholders and the tobacco-processing companies. Each PCS elected a committee from within the community to represent farmers' interests.

Individual farmers form small groups of between three and five members and register with the PCS to become members during a specific production season, normally one season. The main criterion for group membership is based on character assessment done by the farmers themselves. Members give guarantees in relation to the receipt of input credits received from the tobacco-processing companies. In the event of a loan default by any member, other members bear the responsibility for the repayment. As part of the input credit system, a PCS enters into contract with a financial institution to obtain the necessary loans to acquire the required agro-chemical inputs for tobacco cultivation, and thereafter the PCS supplies the inputs to the farmers.

Apart from the distribution of inputs on credit and collecting and selling tobacco leaves on behalf of the members, one of the key roles of the PCS is to receive production quotas from the tobacco-processing companies and allocate the quota to its members. The quota allocation is also used to estimate input needs. The criteria used for quota allocations are the number of curing barns the farmer has, the labour force available for undertaking farm activities and the planting of trees. In addition, the PCSs also serves independent tobacco producers who prefer not to cultivate tobacco under contract farming. The main need of the independent farmers is a market outlet for their produce, which is achieved through sale to the PCS.

A number of operational challenges, including increased side-selling by smallholders, led to the accumulation of debts by the PCSs and declining repayment rates to the two private tobacco-processing companies. As a response to this, the Association of Tanzanian Tobacco Traders (ATTT) – a joint agronomic entity and a joint venture between the two tobacco-processing companies – was formed and mandated as the only supplier of farm inputs on credit, undertaking this task on behalf of the two tobacco-processing companies from

1999. ATTT was charged with a number of activities, including the organisation and administration of tobacco purchasing from the PCSs on behalf of individual companies, in addition to supplying tobacco inputs to PCSs and the management of extension services to growers in the field (interviews with tobacco-processing companies, Morogoro, and ATTT, Tabora, July 2013). This arrangement was seen then as a cost-effective way of organising business (Bureau for Agricultural Consultancy and Advisory Service, 2004).

The contract farming arrangements had a marked impact on tobacco production following a number of steps that were taken. First, some strategic decisions to grant a debt moratorium to farmers were taken. Second, steps were taken to improve debt management at the PCS level. Third, the tobacco-processing companies, through ATTT, had the advantage of controlling the quality of inputs and key farm management practices to improve yields and quality. The fertiliser recommendation was changed, new on-farm handling procedures were implemented and smallholders were encouraged to improve their curing barns (interviews with tobacco-processing companies, Morogoro, July 2013). In addition, improved seeds were provided free by the tobacco-processing companies to ensure that quality seeds were used rather than farmer-kept seeds (Mitchell and Baffes, 2002). In order to reduce waste during transport, the tobacco-processing companies introduced a practice of bale-packing as opposed to the traditional practice whereby smallholders bundle tobacco in jute bags for transport, which results in damage and spillage. As illustrated in Figure 4.1, tobacco production has continued to increase since the early 1960s, though with a major fall in production in the 1998/99 season.

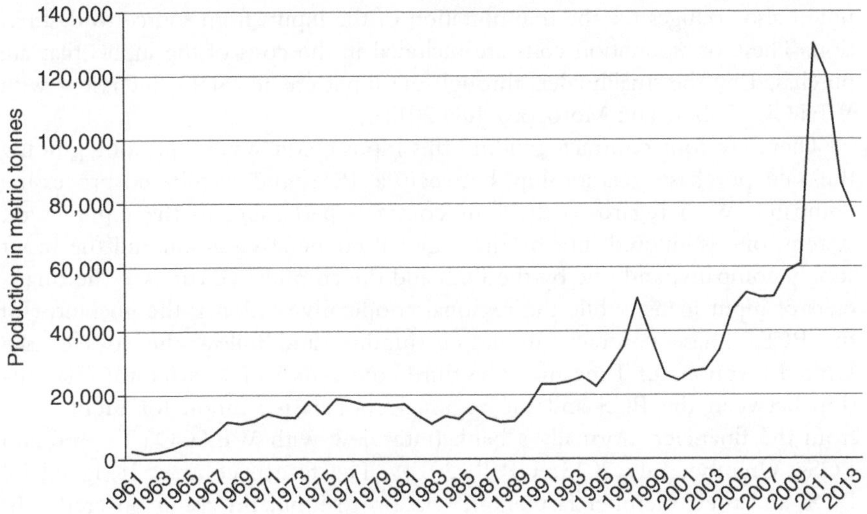

Figure 4.1 Tobacco production, 1961–2013.

Source: FAOSTAT (2014).

Private tobacco-processing companies and regional cooperative unions coordinating contract farming (2009–14)

In 2009, the government reconditioned the contractual arrangements between PCSs and the tobacco-processing companies by re-empowering regional cooperative unions as the prime coordinators of the input delivery system. The tobacco-processing companies were restricted from their involvement in the provision of inputs on credit to smallholders, and the input distribution system was 'transferred' to the regional cooperative unions and their umbrella organisation, the Tanzania Cooperative Apex. This was in line with the mandate provided to them by the Cooperative Act 2003 and the subsequent Cooperative Act (URT, 2013a). The regional cooperative unions represent all the PCSs in a designated geographical area.

Under this new arrangement, the importance of PCSs in the coordination of inputs delivered to smallholders on credit and the collection of tobacco from member farmers remained. In addition, the PCSs remain the key link between the smallholder farmers and the tobacco-processing companies. The supply of inputs involves the farmers as the final recipients and regional cooperative unions as the coordinators of the input delivery system, PCSs, financier and supplying company. Input requirements from the PCSs in a restricted area are aggregated by the regional cooperative union – WETCU in the case of Tabora region – and tenders are issued for the supply of the required inputs. Agro-chemical supplies are procured from those who win the tenders and, in the case of imports (e.g. fertiliser), are transported from Dar es Salaam to the Tabora region. The inputs are delivered to the respective PCSs in accordance with their orders under the supervision of the regional cooperative union. The regional cooperative union also arranges for the transportation of the inputs from source to destination. These transportation costs are included in the costs of the inputs that are purchased by the smallholders through the input credit system (interview with WETCU, Tabora and Morogoro, July 2013).

There are four contracts guiding this input credit system and one guiding the sale–purchase relationship between a PCS and a tobacco-processing company. With regard to the four contracts pertaining to the input credit system, one is entered into by the regional cooperative union and the input supply company, and one by the PCS and the commercial banks as the financiers of input loans, while the regional cooperative union is the guarantor of the PCS. These contracts are legally binding and follow the laws of the United Republic of Tanzania. The third contract set up a guarantor relationship between the PCS and the regional cooperative union for input loans from the financier, normally a bank (interview with WETCO, Tabora, and PCSs, Urambo, July 2013; URT, 2014). The fourth contract is signed by PCSs and their member farmers in relation to input provision on credit. In addition to the four contracts pertaining to the input credit system, a fifth contract entered into by a PCS and a tobacco-processing company specifies the roles and obligations of the respective parties related to the sale and

Table 4.1 International trade in tobacco: export and import, 2010–14 (metric tonnes)

	2010	*2011*	*2012*	*2013*	*2014*
Export					
World total	2,287,122	2,365,499	2,733,715	2,508,196	2,374,953
Brazil	493,003	533,579	624,699	609,927	460,525
China	251,276	223,791	211,433	199,456	159,046
United States	69,263	78,734	165,039	52,251	69,974
Malawi	144,219	159,286	222,106	135,729	256,160
Mozambique	29,527	58,375	55,942	58,938	58,607
Tanzania	69,095	74,022	107,593	69,449	80,347
Zimbabwe	90,196	134,466	131,853	147,873	141,559
Italy	75,706	61,453	76,653	69,414	61,499
Turkey	78,878	68,031	75,680	57,512	67,895
Import					
World total	2,503,282	2,547,118	2,625,167	2,581,828	2,334,994

Source: FAOSTAT (2014).

purchase of tobacco, as well as the obligations of PCSs to ensure that their members act in accordance with good agricultural and labour principles. Extension services are provided by ATTT on behalf of the two tobacco-processing companies (for details, see Chapter 13, this volume).

Meanwhile, the regional cooperative union represents the smallholders (and PCSs) at the sale of tobacco leaves. Staff from the regional cooperative union is present at the market outlet in the villages to mediate in cases of dispute over grades and price settlements. One of the concerns of the smallholders is the allegedly unfair classification and grading of tobacco leaves during marketing of the crop. Currently there are more than 60 grades of tobacco leaves, and the classification system and prices attached to each grade are not clear to most farmers. According to smallholders, this results in lower prices being offered for their produce (interviews with farmers, Urambo, 2012 and 2013).

Evolving chain governance structures and impacts on smallholders

The successful development of smallholder tobacco production can be assigned to 'production under close supervision' (Boesen and Mohele, 1979: 126) and the establishment of well-functioning contract farming arrangements both before and after market liberalisation in the mid-1990s. The contract farming system has been developed under conditions of a relatively high intensity of extension services compared to the production of other crops in Tanzania, as well as close supervision and monitoring of husbandry practices related to the application of agro-chemical inputs and – more recently – labour and environmental practices at the farm level. During the three historical periods examined above, smallholder tobacco-growers' relationships with PCSs or similar institutions (i.e. village cooperatives prior to liberalisation) remained unchanged. Smallholders are linked to PCSs in order to be able to trade their tobacco produce and receive inputs on credit. In addition, the PCSs continued to be linked to the tobacco-processing companies through a captive governance structure. As explained above, in the late 2000s, the power of the tobacco-processing companies versus the PCSs in the input credit system was contested. The former were prevented from involvement in the input credit system by the state, apparently as a means to dissolve the power of companies, who ran the contract farming scheme almost as a monopoly, while the regional cooperative unions assumed the role of coordinating the supply of input credits to smallholders.

This change in the relationship between the tobacco-processing companies and PCSs did not increase the volatility of the relationship with regard to the sale of tobacco. The PCSs remained the main entity between the tobacco smallholders and the tobacco-processing companies and continued to perform a key coordinating role in collecting and trading tobacco to the latter. In addition, with the restructuring of the input credit system, the latter were able to avoid the administrative burden of operating an input credit scheme and tying up working capital (interviews with tobacco-processing companies, Morogoro, July 2013). Through ATTT, the tobacco-processing companies continued to monitor and supervise smallholders regarding their husbandry, labour and environmental practices. Hence to some degree, the two tobacco-processing companies are guaranteed smallholder compliance with key product and process requirements.

On the other hand, with the transfer of the input delivery system to the regional cooperative unions, the PCSs became highly dependent on the unions' capacity to coordinate the input delivery system both in a timely manner and effectively. Apparently, the relationship between PCSs and the regional cooperative union is associated with delayed deliveries of inputs, malpractices and hidden overhead costs that might threaten the sustainability of the PCSs. Delays in supplying inputs to PCSs lead to delays in the distribution of inputs to the farmers, thus jeopardising the quantity and quality of production. As one of the farmers emphasised, contrasting the recent input credit delivery system with the former system:

> With the [tobacco-processing] companies distributing the inputs, they monitored the societies accurately and diligently, something which is not done at the moment. Tobacco production strictly follows a calendar. If input delivery is delayed, it affects farmers in terms of production quality and quantity, and this put farmers into a situation of unrecoverable debts.
>
> (Interview with farmer, Urambo, July 2013)

In cases of late deliveries of inputs, the PCSs are not able to take the input suppliers to court as the input procurement contract is between the input supplier and the regional cooperative union.

Malpractices and hidden overhead costs have apparently also emerged in the wake of transferring the input delivery system to the regional cooperative unions. This issue was raised by almost all the PCSs interviewed in 2013, for example:

> The shift from [tobacco-processing] companies ordering inputs to the present system is even worse! It is normal to pay for unused inputs, and you may find that the cooperative [PCS] is deducted money even for inputs not delivered to it. For example, for the 2012/2013 season, the PCS ordered a total of 5031 bags NPK 10:18:24 fertilizer, but received only 4700 bags, while the record shows the society received 6186 bags! You can see this discrepancy. We also ordered 1400 bags of CAN fertilizer, but we received 750 bags (each bag weighs 50 kg). This trend causes the PCS to use internal sources of income to repay bank debts unnecessarily, and this threatens the sustainability of the Primary Cooperative Societies.
>
> (Interview with PCS, Urambo, July 2013)

Most of the PCSs expressed their concerns with the role of the regional cooperative unions as the prime coordinators of the input supply system. These PCSs preferred instead to become responsible for ordering inputs for tobacco cultivation directly from agro-chemical input providers through tenders.

Furthermore, the PCSs' financial performance has come under pressure due to the increase in side-selling practices by smallholders in recent years. The PCSs are mandated to perform certain functions for their members by the Cooperative Act, including the supply of agricultural inputs and the collection and marketing of produce on behalf of their members (URT, 2013a). For these functions to be performed properly, the capacity of the PCS, including its financial standing, is very critical. The main source of revenue for the PCSs is membership fees deducted from each kilo of tobacco sold (between 50 and 150 Tsh per kg tobacco). The second source is so-called 'agent commissions' paid by the tobacco-processing companies. It is these sources that have to finance PCS operations, including operational and development activities. One of the critical challenges facing the PCSs in recent years is loan defaults by farmers, often caused by side-selling. Side-selling occurs when a tobacco farmer with an input credit contract with a given PCS secretly sells part or all of the harvested tobacco to another PCS using a fellow farmer who is registered to that PCS. The PCS, however, enters into contracts with the banks for input loans guaranteed by the regional cooperative union and has to pay the input loans to the banks irrespective of whether it has fully recovered them from its members. If farmers default on input credit repayments, the PCS has to use its own financial resources to make good the deficit. This also reduces the aggregated revenue obtained from membership fees charged during tobacco sales, as well as its agent's commission.

Six out of the ten PCSs that were interviewed reported that they can only repay bank loans by using their own resources. One of the PCSs recorded a debt of 90 million Tsh arising from side-selling during one marketing season alone (interviews with PCSs, Urambo, July 2016). Side-selling also affects the ability of some PCSs to pay each farmer for the total quantity of tobacco each sells through the PCS, as the PCS is obliged to recover bank loans before cash disbursements to members according to the quantity sold. This threatens not only the sustainability of the PCSs but also the livelihoods of the smallholder farmers. According to the smallholders, side-selling is triggered for several reasons. When a farmer finds that s/he will not meet the estimated production quota for some reason, s/he side-sells in order to obtain at least some cash, rather than recovering (part of) the input credit. Also, the farmers' group arrangement is based on collective responsibility for repaying the input credit, and this apparently also encourages side-selling habits. This is especially the case in circumstances when members realise that there are 'unfaithful' farmers who cause others to repay the loan for them under the collective responsibility agreement. Farmers decide to side-sell their tobacco as a way of avoiding paying for a fellow member in the group. Hence, group reliability with regard to input loans and defaults on the part of some group members seems to foster a (unbreakable?) cycle of side-selling practices.

As explored above, changes in the relationships between PCSs and tobacco-processing companies arose from state-driven initiatives and are not only 'driven' by the strategic needs of the lead firms in the GVC for tobacco, whether the branded international cigarette manufacturers or tobacco-processing companies located in Tanzania. Value chain governance seems to be intertwined with the active role of the state in shaping chain structures and socio-economic outcomes. In the two most recent tobacco-growing seasons (2014–16), there might be signs of further changes to the governance structure as a result of the less concentrated market that has emerged. In addition to the two well-established tobacco-processing companies, a third large international brand cigarette manufacturer, which used to purchase tobacco blends from the two tobacco-processing companies, was allowed to enter the market. This company owns 75 per cent of the shares in the only cigarette-producing company in Tanzania and started sourcing tobacco leaves directly from PCSs in 2014 (Japan Tobacco International, 2016). With the entry of a third company into the market, the two well-established tobacco-processing companies were asked to dissolve ATTT. Whether this development will transform the relationships between the smallholders and the tobacco-processing companies remains to be seen, but the entry of a third company has increased price competition in the market to the benefit of smallholders. In addition, more players in the market allows the PCSs to choose which tobacco-processing company they will enter into a contract with based on 'the package' that the company is prepared to provide directly, such as provision of extension services.

Conclusion

The tobacco value chain in Tanzania presents an interesting case of evolving dynamisms in chain governance structures. This chapter has examined three distinct periods of contract farming and changing relationships between smallholders and tobacco-processing companies. Contract farming was initiated and consolidated by the state during the 1960s. A parastatal coordinated the entire tobacco value chain and the contract farming arrangements between smallholders and parastatal tobacco-processing companies. After liberalisation of the agricultural sector in the mid-1990s, the contract farming scheme was taken over by two private tobacco-processing companies, both subsidiaries of major international tobacco leaf merchants. Like the state-governed contract farming scheme, inputs on credit were distributed to smallholders through PCSs and the costs were deducted at sale to the private tobacco-processing companies. As a means to dissolve the power of the latter in operating the contract farming scheme almost as a monopoly, in 2009 the state prevented their involvement in the input credit system. At the same time, the regional cooperative unions assumed the role of coordinating the input delivery system to smallholders.

As revealed in the analysis, the recent changes in the relationships between the tobacco-processing companies and the PCSs appear not to have had any direct negative consequences for the tobacco companies' ability to fulfil forward contracts with the international cigarette companies. To some degree, they still have a reliable supply base through contract farming, though indirectly, a lack of efficiency in the input delivery system might jeopardise their tobacco uptake in Tanzania.

The PCSs, however, have become more vulnerable despite the aim of 'empowering' smallholders in relation to the tobacco-processing companies in the input credit system. The PCSs still perform a crucial role in coordinating input credit to smallholders and as such are vulnerable to inadequate coordination of the input supply system. This, along with increasing debt accumulation as a result of side-selling practices by smallholders, might threaten the sustainability of the PCSs. Delays in inputs distributed to farmers, as well as malpractices related to the quantities of inputs distributed by the regional cooperative unions to the PCSs, result in additional costs to be carried by the PCSs. Since inputs are central to tobacco production, there is a need to strengthen the capacity of the regional cooperative union to perform this role efficiently and effectively. Alternatively, direct provision of inputs on credit by the tobacco-processing companies has the advantage of maintaining the contractual relationship between farmers and buyers, and it might contribute to the more efficient coordination of the entire chain.

Tanzania's apparent economic dependence on tobacco and its readiness to adopt strict tobacco control policies in line with global conventions (e.g. WHO) are not simple factors. While seen as a success story in terms of providing a livelihood for numerous smallholder farmers, as well as contributing to national economic development through foreign exchange earnings, tobacco production in Tanzania and the chain as a whole face resounding challenges that pose a significant dilemma. Strict tobacco control procedures in accordance with the WHO protocol have been implemented to reduce tobacco use and reverse the increasing number of young tobacco users in the country. As a result of increased tobacco consumption, treatment of and expenditure on tobacco-related health problems are mounting. Long-term planning will be required to find alternative sources of livelihood for smallholders in rural parts of Tanzania, as well as alternative foreign exchange earners for the country as a whole.

Notes

1 For more details on the development of the smallholder tobacco sector from the early 1960s, see, for example, Boesen and Mohele (1979) and Ponte (2002).
2 For a discussion of governance typologies and chain coordination, see Chapter 1, this volume; also Gibbon and Ponte (2005).

References

Agricultural Marketing Resource Centre (2012). *Tobacco Profile*. Online: www.agmrc.org/commodities-products/specialty-crops/tobacco-profile. Accessed 16 September 2016.

Bank of Tanzania (2016). *Quarterly Economic Bulletin*. December 2015. Online: www.bot.go.tz/Publications/publicationsAndStatistics.asp.

Boesen, J. and Mohele, A. T. (1979). *The 'Success Story' of Peasant Tobacco Production in Tanzania*. Uppsala: Institute of African Studies.

Bureau for Agricultural Consultancy and Advisory Service (2004). *Institutional Mapping of Tobacco*. Unpublished Consultancy Report. Sokoine University of Agriculture, Tanzania.

FAOSTAT (2014). Food and Agriculture Organization of the United Nations. Online: www.fao.org/faostat/en/#data/QC. Accessed 26 August 2016.

Fold, N. (2002). Lead Firms and Competition in 'Bi-polar' Commodity Chains: Grinders and Branders in the Global Cocoa-Chocolate Industry. *Journal of Agrarian Change*, 2(2), 228–247.

Gereffi, G. (1994). The Organization of Buyer-Driven Global Commodity Chains: How US Retailers Shape Overseas Production Networks. In G. Gereffi and M. Korzeniewicz (eds), *Commodity Chains and Global Capitalism* (pp. 93–123). Westport, CT: Greenwood Press.

Gereffi, G. (1999). International Trade and Industrial Up-grading in the Apparel Commodity Chain. *Journal of International Economics*, 48, 37–70.

Gereffi, G., Humphrey, J. and Sturgeon, T. (2005). The Governance of Global Value Chains. *Review of International Political Economy*, 48(1), 78–104.

Gibbon, P. and Ponte, S. (2005). *Trading Down: Africa, Value Chains & the Global Economy*. Philadelphia, PA: Temple University Press.

Humphrey, J. and Schmitz, H. (2002). How Does Insertion in Global Value Chains Affect Upgrading in Industrial Clusters? *Regional Studies*, 36(9), 1017–1027.

Ilembo, B. (2015). Contract Farming and the Dynamics of Tobacco Value Chain in Tanzania. Unpublished PhD Thesis. Mzumbe University, Tanzania.

Japan Tobacco International (2016). Online: www.jti.com/brands. Accessed 26 August 2016.

Joossens, L. and Gilmore, A. B. (2013). The Transnational Tobacco Companies Strategy to Promote Codentify, Their Inadequate Tracking and Tracing Standard. *Tobacco Control*, OnlineFirst. http://tobaccocontrol.bmj.com/content/early/2013/03/11/tobacco. Accessed 19 September 2016.

Li, D. X. and Guindon, G. E. (2013). Income, Income Inequality and Youth Smoking in Low and Middle Income Countries. *Addiction*, 108(4), 799–808.

Mitchell, D. and Baffes, J. (2002). *Tanzanian Agricultural Export: Challenges and Constraints in a Global Economy*. Washington, DC: World Bank.

Neilson, J. and Pritchard, B. (2009). *Value Chain Struggles: Institutions and Governance in the Plantation Districts of South India*. Chichester: Wiley-Blackwell.

Ponte, S. (2002). *Farmers and Markets in Tanzania*. Oxford: James Currey.

Sturgeon, T. (2002). Modular Manufacturing Networks: A New American Model of Industrial Organisation. *Industrial and Corporate Change*, 11(3), 451–496.

URT (2013a). *The Cooperative Act*. United Republic of Tanzania. Online: www.kilimo.go.tz/legislation and regulations/legislations/The Cooperative Societies Act 2013(1).pdf. Accessed 21 June 2016.

URT (2013b). *The Economic Survey 2012*. Ministry of Finance, United Republic of Tanzania. Online: www.mof.go.tz/mofdocs/Micro/2012/ECONOMIC SURVEY BOOK 2012.pdf. Accessed 25 June 2016.

URT (2014). Hotuba ya Waziri wa Kilimo, Chakula na Ushirika Mheshimiwa Stephen Masato Wasira (MB) kuhusu Makadirio ya Mapato na Matumizi ya Fedha ya Wizara ya Kilimo, Chakula na Ushirika kwa Mwaka 2015/2016. United Republic of Tanzania. Online: www.kilimo.go.tz/speeches/budget speeches/HOTUBAYA BAJETI-KILIMO.pdf. Accessed 21 June 2016.

URT (2016). *National Five Year Development Plan, 2016/17–2020/21*. United Republic of Tanzania. Online: www.nationalplanningcycles.org/sites/default/files/planning_cycle_repository/tanzania/fydp2_ii_april_1.pdf. Accessed 25 July 2016.

Warner, K. E. and Mackay, J. (2006). The Global Tobacco Disease Pandemic: Nature, Causes, and Cures. *Global Public Health*, 1(1), 65–86.

WHO (2003). *WHO Framework Convention on Tobacco Control*. Geneva: World Health Organization. Online: www.who.int/tobacco/framework/WHO_FCTC_english.pdf. Accessed 21 June 2016.

5 Successes and barriers regarding small and medium-size enterprises (SMEs) in the value chain for sunflower in Tanzania

Does contract farming reduce value chain coordination problems for SMEs?

Daniel Mpeta, Joseph Andrew Kuzilwa, Batimo Sebyiga and Niels Fold

Introduction

Many studies analysing the effects of contracts on crop value chain development tend to focus on actors in upstream nodes, examining, for example, whether and how involvement in contract farming has improved the efficiency or incomes of participating farmers, in particular smallholders. Few studies analyse the individual capacity and strength of small-scale downstream actors, for example, the small or medium-size agro-industrial firms that actually provide the contract. However, in an economic environment of contracting smallholder farmers for the commercial production of a given crop, the ability of small and medium enterprises (SMEs) to create and sustain contract farming operations is essential. Successful contract farming largely depends on the extent to which a firm is able to coordinate and control its supply and demand chain while keeping up with competitive pressures in the industry.

The significance of the private sector in the development of the sunflower value chain in Tanzania cannot be over-emphasised. A growing number of privately owned, small-scale oil mills have sprung up to crush sunflower seed in many parts of the country. For instance, there are about 47 small-scale oil-processing mills registered with the Central Zone Sunflower Oil Processing Association (CEZOSOPA) alone. Ideally, the growth of these small-scale enterprises provides sunflower producers with local market outlets for their crop and helps ensure the long-term viability of all the activities related to the sunflower value chain in the region.

Tanzania has 26 major manufacturing plants producing and refining edible oil. Large-scale processing is confined to plants in Dar es Salaam, Morogoro, Arusha and Mwanza. Production levels in processing are significantly lower than installed capacity because of insufficient supplies of edible oil seeds. Currently, national production of both large- and small-scale extracted oils contributes about 40 per cent of the national edible oil requirement (URT, 2016). The vegetable oil industry is dominated by six large-scale firms with well-known branded products, two of which are located in the central and northern zones. The total processing capacity of large firms is not exactly known, but it is estimated at more than 900,000 metric tonnes per year. Unlike the SMEs, which concentrate on processing a single vegetable crop like sunflower mainly for the domestic market, the large firms process multiple crops like sesame, soya and groundnuts. The latter firms target both the domestic and international markets for their inputs (including imports of crude oil from Asia) and outputs. For instance, BIDCO (T), which acquired Unilever's businesses in East Africa in the 1990s, is exporting oil to the COMESA market, as well as to markets in the European Union. While large firms compete with SMEs for supplies of sunflower seeds from smallholder farmers, they also provide a useful market for oil cake from SMEs: further oil is extracted, and the resulting oil cake is well suited for use as animal feed.

The common and major problem facing SMEs involved in sunflower oil production is the lack of an adequate supply of quality oil seed to enable optimal and continuous processing all year round. To overcome this problem, SMEs have attempted to set up contract farming systems in order to organise production so that large and steady quantities of quality seed are produced and secured. For a couple of years now, contract farming has been tried in the central agricultural zone of Tanzania, and the number of farmers and firms entering into contracts has risen. The geographical coverage of this study includes the administrative regions of Dodoma and Singida, where sunflower is widely grown. The area contributes significantly to the total national output of sunflower, of which it produces about 60 per cent (RLDC, 2012). These regions have a high concentration of small and medium-size sunflower oil-processing plants in urban centres and townships, offering a potential market for sunflower farmers in the area. And, contract farming among SMEs is significant.

Observers argue that the involvement of private firms (processors) is likely to increase both productivity and total volumes of sunflower produced in the area (RLDC, 2009), while also improving the overall economic conditions of smallholders. This study documents the crucial importance of elements 'external' to the contracts for the successful performance of contractual systems, that is, systems from which both smallholders and the small-scale oil-processing firms benefit economically and where the transactions between the parties are maintained and repeated. These are elements like local seed multiplication and distribution projects, use of demonstration plots, use of mobile phone text messages (for the communication and dissemination of important issues related to crop production), collaboration with the local government authorities (LGAs) and possibly the increased involvement of female farmers. The presence of some or all of these elements is the main reason for the reduction or even elimination of the major problem for the SMEs involved, namely side-selling to other buyers under spot-market conditions (see the subsequent sections).

This chapter is based on the findings of qualitative in-depth interviews with seven small-scale sunflower oil-processing firms, including six processors who were still operating with contracts and one who had stopped contract farming in the past year.[1] Besides the 'external' elements discussed in this chapter, the SMEs in sunflower oil-processing face numerous other challenges, like limited access to state-of-the-art technology, a lack of access to credit and investment capital, high-cost and unstable supplies of electricity and various marketing constraints (logistics, product standards, etc.). However, these issues are of a more generic nature and are not directly related to the contractual system, hence they are not discussed further in this chapter (but see Mpeta, 2015, for details).

Coordination in smallholder-dominated agro-industrial value chains

Contract farming is held to be an essential feature of modern agricultural chains (FAO, 2011, 2012; UNIDO, 2011). The establishment of contracts between producers and businesses has been found to provide an effective link contributing to the modernisation and improvement of the functioning of value chains. Through contracts, processors and other businesses are in principle assured a steady supply of the right quantity and quality of raw materials from producers and, in their turn, farmers are assured of reliable and profitable market outlets. Formal contracts specify production and purchasing procedures, as well as credit and pricing mechanisms. Generally, producers provide land and labour, while buyers provide specialised planting materials such as improved seeds and other production tools (Raynolds, 2002). The popular assumption is that both firms and farmers engaged in contract farming inevitably gain equally from the potential benefits associated with contractual arrangements.

However, for a firm to achieve its goal of driving a chain based on its market power, it needs to develop necessary competences in tasks such as technology acquisition, finance, business management and marketing so that it has the ability to facilitate activities such as the provision of extension services, training, information and assistance in physical needs (e.g. inputs and other farmers' needs, like storage facilities) (WEF, 2010). Above all, for it to carry out such contractual activities successfully, it should be able to develop and manage strategic relationships with farmers so that activities can be effectively coordinated.

Value chain coordination by agribusiness firms refers to all the strategies a firm needs to ensure that there are value additions in both forward and backward linkages of the chain. Value chain coordination at the firm level binds together different management activities both up- and downstream to bring about inter-company relationships that enable the firm to deliver marketable products at as low a cost as possible. Ideally the firm increases its competitiveness, wins a larger market share and earns higher profits (UNIDO, 2009). The value chain perspective is mainly rooted in the work of Porter (1985), whose value chain approach focuses on supply chain management by firms, essentially a matter of a firm's management of operations in value chains within a national territory. In the more recent value chain literature (Gereffi, 1999; Gereffi *et al.*, 2005), the value chain approach focuses on different types of governance structure in which the coordination of global economic activities is made possible through non-market relations and means.

The so-called Value-Chain Approach to Pro-Poor Economic Development (Hawkes and Ruel, 2011) is a market-based economic development framework that presupposes the development and welfare of the poor through the development of a private sector in those activities and sectors that have a high concentration of the poor, such as smallholders in agriculture and small-scale firms in rural parts of developing countries (Hawkes and Ruel, 2011). It is argued that, given free market forces, SMEs in developing countries have to compete with large agribusinesses in developed economies. However, given their weak economic backgrounds, they are easily outcompeted and therefore do not play a role in facilitating poverty reduction.

The Value-Chain Approach to Pro-Poor Economic Development was developed to encourage participation of the poor by establishing the competitiveness of private firms in those sectors where the poor who are involved have some kind of comparative advantage. The poor need support from private enterprise in order to participate effectively and improve their positions in the value chains in which they are engaged. Such an approach would enhance equitable economic growth, as the poor would find opportunities to generate benefits for themselves just as enterprises do.

In general terms, value chain development also requires concerted efforts involving both private and public parties. The performance of private firms, and of SMEs in particular, depends on the availability of critical support services such as rural roads, electricity and information services, management support and financial facilities, all of which are best provided by the government (UNIDO, 2011). A lack of or inefficiency in critical support services creates problems in the logistics of production and marketing and makes private investment costly. Furthermore, the government needs to create an enabling economic environment by setting up regulatory mechanisms and policies that provide incentives for investors, including access to credit facilities at a reasonable cost (WEF, 2010).

Sunflower production in Tanzania: growth and regulatory framework

Sunflower in Tanzania was for a long time merely a traditional crop without any significant commercial value. More recently, however, sunflower has become important because of its growing monetary value, which is associated with its relatively low cost of production. The local demand for sunflower oil is increasing, and its potential to become a major agricultural export commodity is also rising (ARI Ilonga, 2008; RLDC, 2008). The total amount produced has steadily increased over time, from an average of 80,000 tons per year in 2000/01 to about 2,625,000 tons per year in 2015/16 (URT, 2015). The total amount of oilseed that is domestically produced comprises sunflower (35 per cent), groundnuts (25 per cent), sesame (13 per cent), seed cotton (19 per cent), cotton seed (7 per cent) and other (palm oil) (1 per cent) (URT, 2016).

Tanzania still imports edible oil: it is estimated that about 60 per cent of total edible oil consumed in the country is imported, while only 40 per cent is produced domestically (ARI Ilonga, 2008). This suggests that local producers have a huge potential market while at the same time facing a stiff competitive challenge from imported edible oil. To a certain extent, the policy environment protects local seed and oil production by restricting oil imports. Since 2010 a 10 per cent import duty has been imposed on refined and semi-refined imported edible oil, and the government exempts all agro-processing industry firms that source 70 per cent or more of their raw materials from the local market from value added tax (VAT). In return, the government requires a commitment from agro-processing firms that they increase production every year (MMML, 2010).

Another agricultural and trade policy that supports the development of the oilseed industry includes the Sustainable Industrial Development Policy, which is anchored within the national Agricultural Sector Development Strategy and focuses on the development of agro-processing industries (URT/FAO, 2008). The strategy seeks to create an inspiring economic and legal environment in which the private sector can thrive. It also seeks to encourage public–private collaboration whereby the government commits itself to making the necessary investments in basic public agricultural services, while the farmers invest in agricultural production. The regulation of the edible oil sector in Tanzania also includes strict standards of production overseen by the Tanzania Bureau of Standards (TBS) (URT, 2002) and quality regulation overseen by the Tanzania Food and Drugs Authority (TFDA) (URT, 2003a).

Seed production and certification in Tanzania is governed by the Seed Production Act 2003, as amended in 2007 and administered by the Ministry of Agriculture, Food Security and Cooperatives (RLDC, 2008). There are also institutions that play specific roles in the development of seeds for the sunflower value chain. This includes the Tanzania Agricultural Seed Agency (ASA), whose main responsibility is producing and supplying high-quality agricultural seeds on a sustainable basis. The major functions of this government agency include (URT, 2003b):

i To expand seed production and distribution networks so as to facilitate access to seed for farmers.

- ii To promote increased private-sector participation in seed-industry development through the establishment of public–private partnerships or joint ventures in seed production and distribution.
- iii To promote increased demand for certified seed varieties from farmers.
- iv To strengthen research capacities for breeding and producing varieties that address farmers' specific demands.

Another institution is the Tanzania Official Seed Certification Institute (TOSCI), which is responsible for seed certification. Moreover, there are also agricultural research institutes like Ilonga and the Sokoine University of Agriculture that breed new and improved seeds.

In the Central Zone, an important role is played by a private-sector institution, the regionally embedded Rural Livelihood Development Company (RLDC), a non-profit organisation funded by the Swiss Agency for Development. Its main activities are (i) to work with other stakeholders in the sector (e.g. local government authorities, banks, farmers' organisations) and address the production and marketing constraints facing sunflower farmers and processors; (ii) to facilitate the formation of alliances between farmers, processors and other stakeholders who may be interested in the sector's development; and (iii) to propose convincing business development models such as contract farming systems and local seed multiplication projects (to improve seed availability at the farm level) and enhance the production and productivity of sunflower.

Characteristics of the local SMEs engaged in sunflower oil-processing

All seven local oil-processing firms that are included in this study are family-run businesses. According to the national definition of small businesses (URT, 2002), they fall under the small enterprise category on account of the number of workers they employ, which is fewer than 50 in each case (see Table 5.1). The age of these SMEs indicates that there is an almost continuous inflow of new entrants into the sub-sector and that the barriers to entry are relatively low. Most of these SMEs have been in business for less than ten years, which is in line with the fact that the sunflower sub-sector in Tanzania is a relatively new sector that is undergoing gradual but significant commercialisation.

However (as indicated in Table 5.1), none of the firms are able to operate for all the months in a year. The average time of operation is currently eight months, implying that there are on average about four months in the year when plants are idle.

The most common reasons for the low capacity utilisation are (1) limited working capital and (2) an inadequate supply of raw materials due to low total seed production (low supply). First, even in areas where local production is sufficient to support prolonged seed availability, some firms cannot buy enough for the whole annual production cycle due to a lack of working

Table 5.1 Characteristics of local sunflower oil processing firms*

Codes	*Firm A*	*Firm B*	*Firm C*	*Firm D*	*Firm E*	*Firm F*	*Firm G*
Ownership	Family	Family	Family	Family	Family	Family	Family
Age of firm in years	6	3	5	10	13	5	8
Workers							
Permanent	8	9	8	4	33	11	6
Temporary	–	–	–	12	–	–	–
Procurement channels							
Contract (%)	100	95	80	69	5	0	15
Spot-market (%)	0	5	20	31	95	100	85
Utilisation rate							
Months per year	10	4	10	6	10	8	7

Source: authors' fieldwork (2012).

Note

* Firm A: Songela Inv. Co. Ltd; Firm B: Ringo Consolidated Co. Ltd; Firm C: Three Sister Oil Mills Co. Ltd; Firm D: Rig Investment; Firm E: UncleMilo; Firm F: Furaha Dodoma Oil Mill; Firm G: Jackma J Enterprises.

capital. For instance, firm A is located in Iramba District, where the quantity of total sunflower produced by contract farmers is more than what the firm can absorb in a batch or a few batches of procurement. However, the breakdowns in the production process for this firm are not due to a lack of raw seed but to a lack of the financial resources to purchase supplies for a whole year.

Second, the seed-processors cannot obtain the whole quantity of raw materials they need because the supply of the material is insufficient. Most of the SMEs rely on both spot-market purchases and contract supplies for their provisioning of raw materials (see Table 5.1). Interviews confirmed that the main driving force behind the contract farming strategy is the need for a consistent supply of standard-quality sunflower seeds. In theory, among other things contract farming may help ensure coordinated actions: contracts can positively influence production and productivity through contractual arrangements that ensure the timely availability and delivery of production inputs such as seed and a market outlet for the produce. Thus, by setting up contracts, firms strive to improve the coordination of production (Bogetoft and Olesen, 2004; Singh, 2002).

However, the record of SMEs involved in the contract farming of sunflower seed is mixed when it comes to the performance of contracts. Some firms can be considered successful in their contract farming in that they assess the percentage at which they have been able to secure a supply of raw materials by way of contracts (see Table 5.1). Firms A and B source 100 per cent and 95 per cent respectively of their total raw material needs through contractual arrangements, while firms C and D rely on contracts for a somewhat smaller share of the quantities needed. In contrast, firms E, F and G have weak contracts expressed by low contributions, namely between 5 per cent and 15 per cent of total raw material sourced from contractual arrangements. Firm F indicates a completely failed contractual scheme, with none of its raw material needs sourced via contracts. Hence, having established contractual arrangements as such does not guarantee a sufficient and certain supply of raw material. This implies that contracts work but are subject to other factors. The key issue of interest is why some SMEs are relatively more successful than others with respect to the implementation of contractual arrangements. The next section analyses similarities and differences in the content and form of the contractual arrangements implemented by the SMEs in this sector.

Characteristics of the contracts between SMEs and smallholders

The contracts we have been discussing are similar in many ways, not least due to the fact that they are all based on a contract farming model facilitated by the RLDC (see above). First, contracts are only valid for one year or one crop season. Second, the main issue covered by the contracts is the provision of inputs, in particular seed, to ensure that farmers use quality seed so that they produce quality output in increased quantities. Hence, seed provision is implemented by all the SMEs (see Table 5.2).

Other services, like tractor services, are rarely provided to farmers: only firm C does this, but even then only to a very few farmers who were identified as being too poor to afford to buy tractor services on their own. The cost of tractor services for one acre was a fixed rate of 25,000 Tsh, which the farmer had to pay back in seed after the harvest. There are no contracts that include other inputs like fertilisers or pesticides that may be required to enhance farmers' productivity. In any case, farmers are able to grow sunflower in this part of the country without requiring other inputs apart from seed. This suggests that soil fertility and other environmental conditions are basically very supportive of sunflower seed production, but this may not remain the case in the future. In places like Mpwapwa there are growing problems with pests (e.g. worms, grasshoppers, birds) and diseases, which in the near future may require increased use of pesticides (chairperson of TASUPA, personal communication).

Third, the firms are trying to promote the participation of more farmers in the schemes to increase total output. The number of small-scale farmers involved varies from one firm to another. Some firms (B and C) were able to contract as many as 3,000 farmers in a single season in an area of up to 25 villages in Mpwapwa and Kondoa Districts respectively (see Table 5.2).

Fourth, all contracts are informal in the sense that they are designed and implemented without strictly observing the legal procedures. Farmers are only presented with written copies of contracts once, during the initial negotiations at the beginning of the farming season, and they are not given a copy

Table 5.2 Characteristics of contractual arrangements

Chain processes	*Firm A*	*Firm B*	*Firm C*	*Firm D*	*Firm E*	*Firm F*	*Firm G*
Procurement							
By contracts (%)	100	95	80	69	5	0	15
Spot-markets (%)	0	5	20	31	95	100	85
Inputs provided to farmers							
Seeds	Yes	Yes	Yes	Yes	Yes	Yes	Yes
Extension service	Yes	Yes	Yes	Yes	Yes	Yes	Yes
Information	Yes	Yes	Yes	Yes	Yes	No	Yes
Tractor service	No	No	Yes	No	No	No	Yes
Price determination	Market	Market	Market	Market	Market	Market	Market
Assistance from RLDC							
Financial	Yes	Yes	Yes	Yes	Yes	Yes	Yes
Technical	Yes	Yes	Yes	Yes	Yes	Yes	Yes
Collaboration with LGA	Yes	Yes	Yes	Yes	No	No	No
Prod. strategies							
Seed multiplication	Yes	No	Yes	No	No	No	No
FF school	No	Yes	Yes	No	No	No	No
Use of SMS	Yes	No	No	No	No	No	No
Warehouse	No	No	No	Yes	No	No	No
Tools (canvas)	Yes	Yes	Yes	Yes	Yes	Yes	Yes
Farmers contracted	Over 800	3,000	3,000	Over 500	Over 600	–	2,450
Female farmers representation	Over 63%	49%	Over 70%	Not specified	Not specified	Not specified	Not specified
Major problems of CF							
Limited capital	Yes	Yes	Yes	Yes	No	No	No
Side-selling	No	Yes	Yes	Yes	Yes	Yes	Yes

Source: authors' fieldwork (2012).

except for the papers they have signed to acknowledge receipt of the inputs supplied. Implementation of the contracts therefore relies very much on personal relationships between the owner of the firm and the participating farmers, and trust is a major factor driving compliance. Technically, the current contracts are best described as 'one-sided' because of the excessive dominance of the SMEs, particularly in drawing up the contract.

Finally, unlike most standard agro-industrial contracts, the sunflower contracts make no provision for forward pricing; price is left to be determined by the market, as transactions are based on the current market price.

Turning to the differences between contracts, the most notable and important one is in how seeds are distributed. Some firms distribute seed to their farmers on credit at the retail price, while others distribute on a basis of both cash (often at below the retail price) and credit. Firm F is the only firm that distributes seed wholly on credit. Three firms (C, D and E) charge the base market price[2] for the seed they purchase from the seed-producing agency (ASA). Firms A and B had their own seed multiplication projects in which the seed produced was sold to the contracted farmers at a considerably lower price (see below for details).

The modality to be used in seed distribution depends on the firm's own assessment of the level of trust between the firm and the contracted farmers. Interviews with SME managers revealed that there is a high risk associated with distributing seed on credit because of the widespread practice of side-selling, a common problem for all firms (as indicated in Table 5.2 and discussed below). Hence, seed was often distributed on credit, but little or no output at all was delivered to the company after the harvest. Firms E and D, operating in Kongwa District, have experienced severe problems with side-selling and as a consequence have stopped distributing seed on credit. Thus, while distributing seed on credit is intended to give farmers access to quality seed (thereby expanding and improving their production, despite their liquidity constraints), distributing seed on a cash basis was actually a way of facilitating the availability of seed to farmers in a manner that minimises risk.

The role of 'external elements' for successful contractual arrangements

Although contracts have some similarities and fewer differences, as explained above, the overall outcomes are profoundly different: some SMEs are reckoned a success, while others complain of unsatisfactory results. In this section, the factors attributed to differences in the performance of contracts between the two groups of SMEs are identified and discussed.

The better-performing firms (here interpreted as firms A, B, C and D – see Table 5.2) are similar to other firms in terms of what they provide to their contracted farmers and the general contract conditions. They all provide quality seeds, facilitate extension services and provide tools for harvesting, such as seed cleaners and canvases to help farmers maintain clean and high-quality produce. Thus, variations in the performance of contracts seem to be related to strategic practices that are specifically undertaken by these firms on the basis of individual initiatives. These interventions are somehow 'external' to the contracts but have been demonstrated to have a significant impact on the performance of contractual agreements. These elements are described and examined in the remaining part of this section.

Seed multiplication

The problem of the low availability of improved seeds has always been of great concern to smallholder farmers: poor-quality seed is a major cause of low productivity and low income from sunflower production for smallholder farmers. Successful contracting firms are those that have managed to provide a solution to this basic problem by devising new and innovative ways of having seed multiplied and distributed locally in a manner that is easy and cost-effective.

In partnership with the RLDC and the respective district authorities of Iramba and Mpwapwa, firms A and B have managed to mobilise resources and farmers to undertake their own local production of improved seed. This improved sunflower seed is multiplied from a top-quality seed ('certified one') purchased from the ASA. The improved seed has a higher germination rate and a high oil content of 40 per cent while local seed is believed to have an oil content of between 25 and 35 per cent. After thorough inspection by the Official Seed Certification Institute (TOSCI), the locally produced sunflower seed is declared a quality seed (QDS) and is released for distribution to farmers.

This QDS is distributed to farmers at a price that is half that of the seed from ASA or other agricultural research centres that is sold by stockists. The main results of this successful intervention in easing the problem of low seed availability are twofold. First, farmers involved in seed multiplication activities benefit directly from higher farm gate incomes, as one bag of quality seed earns a farmer on average a revenue of 105,000 Tsh per bag, a far higher revenue than the normal mean of 35,000 Tsh per bag for 'traditional' sunflower seed.[3] Second, local production of sunflower seed has boosted overall accessibility to seed input. Farmers can obtain sufficient seed of acceptable quality at a reasonable price of 1,500 Tsh per kg, as opposed to the expensive seed ('certified one') from the ASA at 3,500 Tsh per kg.

In response to this situation, many farmers have joined the contract scheme to produce sunflower seed. Yield has more than doubled from a low of three bags to eight to ten bags on average per acre. In addition, the quality of the yield and the overall volume of quality produce in the area have increased, which explains the success of the contract schemes for firms A and B.

Use of Farmer Field Schools (FFS)

Another pronounced problem constraining sunflower farming is the lack of proper crop-husbandry skills among farmers. Demonstration plots offer a solution to this problem in providing opportunities for the farmers to learn by being able to view the right practices for sunflower production and thereafter replicating them on their own farms at the right time. In all places where demonstration plots are used, farmers have improved their agricultural skills, enabling them to increase productivity and overall total output.

Firms B and C consider the use of demonstration plots to be the most important factor explaining the significant differences in the performance of their contracts. Demonstration plots are useful because they enable a farmer to emulate what is done and how it is done: for example, farmers are shown how to maximise the use of farmed land by planting in lines at recommended intervals (75 cm between pits and 30 cm apart with a depth of 5 cm) instead of the traditional method of simply broadcasting seeds in the field, thus enabling farmers to optimise the use of their land and inputs like seed. This means that only 3–4 kg of seed are used for each acre, and the returns per acre are relatively high. Given the generally low level of education among the rural population, practical training on demonstration plots appears to be an ideal way to induct farmers with the necessary knowledge and skills pertaining to sunflower production.

Support from local government authorities (LGAs)

Firms A and B point to the role of local government (the district council and village leaders) and express their appreciation of it, as they consider it to be an important factor in the success of their contract farming projects. The type of support given can be divided into two categories. One is support to raise farmers' awareness when village leaders are actively engaged in mobilising people for meetings in villages between contracting firms and villagers. Such efforts positively influence villagers' choices to participate in contracts. The other is the support given by the district government to private enterprises' specific projects in providing extension services and personnel to farmers.

In Iramba and Mpwapwa Districts, where firms A and B operate, extension officers from the district agricultural development office were assigned to instruct and monitor seed multiplication projects designed to produce quality seeds. These services, which otherwise would be an expensive responsibility for the contracting firms, were covered financially by government, which really boosted the firms' contract farming schemes. The positive response from the village and district authorities significantly contributed to the development of community goodwill among farmers towards the firms in these areas: watching their leaders collaborating with the firms strengthened the farmers' trust in the latter and increased their loyalty to them.

This partly explains why defaulting on loan payments and side-selling were not problems for firm A, were insignificant for firm B and relatively low for firm C compared to the other firms, whose defaulting rates threatened the survival and sustainability of their contract schemes. The owners of firms A and B acknowledged that they benefited from collaboration with the district and village authorities, to which they attributed the success of their contract farming schemes. Hence, government support may be a crucial factor in the success of contract farming projects.

Greater female representation in contracts

Managers of firms A and B considered the success of their contracts to be partly due to the inclusion of more female farmers in them, as they consider women to be more honest and more fully committed to the terms of the contract, which they do not cheat or abuse, but tend to observe stringently so as to reach their set goals. However, the managers of these firms are themselves women. It is hard to tell with certainty whether the inclusion of more women farmers was an objective choice or just a convenient choice for these managers because they preferred working with their fellow women. The revelation, however, points to the fact that the gender aspects of contract formulation and implementation probably need to be considered. This outcome suggests that, if it were explored further, women might be found to constitute an untapped resource that may be a defining factor in the successful performance of contract farming.

For instance, the greater inclusion of women has enabled the firms to avoid side-selling, a problem that has widely affected all the other firms: with 63 per cent and 70 per cent rates of female involvement in contract farming respectively, firms A and B enjoy a high degree of business reliability, as manifested in the absence of side-selling. Transaction costs related to open market purchases are avoided, as the dependence of the two firms on spot markets are as low as 0 per cent and 5 per cent (Table 5.1).

Use of mobile phones

This is an invention of firm A alone. In partnership with companies that provide mobile phone services in the country (Vodacom-Tanzania, Tigo and Airtel), the firm has established a system whereby all the contracted farmers can be reached simultaneously through their mobile phones, using short text messages to instruct or remind them of, for example, the feasible starting time of a certain farm activity (such as planting, weeding, etc.) according to the season or stage of plant development. Short messages of this sort have been a useful tool for firm A in coordinating sunflower production by farmers, and this fits well with the expanding government scheme to bring telecommunication services to rural areas and remote parts of the country. Using this approach, firm A has been able to influence the farmers' behaviour effectively by ensuring timely training in sunflower agronomics in the area. Twice in a farming season, the firm provides training and advice on both proper crop husbandry (in November) and the proper post-harvest management of the produce (in May). By means of these measures, firm A has improved sunflower contract farming practices to such an extent that it obtains all its raw seed supply from contracted farmers.

Side-selling: the main barrier to contractual arrangements

For those firms that experienced a relatively bad performance on the part of the contractual arrangements (firms E and G) or complete failure (firm F), the most pressing problem was side-selling, which the SMEs considered a clear breach of contractual obligations by the participating farmers. All firms except firm A admitted that this is a critical issue in the management of contracts. The factors contributing to side-selling seem to be related to how contracts are agreed over issues of mutual interest, such as price. As mentioned above, the current contracts are different from standard contracts in that they do not spell out the price or give any calculation formula for sunflower seed: price is left to be determined by the market when the selling season begins. In effect this contractual arrangement appears to encourage side-selling and other opportunistic behaviour by farmers. There are many buyers, including those who do not have any previous agreement with farmers, who flock to the villages to buy sunflower produce for cash. This situation leads to sudden increases in procurement competition, and prices usually increase radically, tempting farmers to disregard their contracts.

For instance, firm G could not buy much from the contracted farmers in Bahi District during the 2012/13 harvest season because of a rapid unexpected rise in price. Competition had intensified nationally and attracted big edible-oil companies like Murzah Oil Company from Dar es Salaam, who raised price offers to 55,000 Tsh per bag against the normal spot-market price of 40,000 Tsh, and most local buyers (including the contracting firm G) were crowded out. Similarly, in Kongwa District, firms D and E experienced stiff competition from free traders from Kibaigwa and from as far away as Dar es Salaam. Firm H could not buy anything from its contracted farmers in Bahi District[4] due to a sudden disagreement over interpretation of the 'market price' clause, which had nonetheless been agreed previously. The 'market price' agreed in the contract was interpreted by farmers as the market price in Dodoma town, not the price in the villages in Bahi District.[5]

Hence, a contractual setting in which agreements are made on the terms of exchange but which excludes price appears to have an in-built danger of side-selling. On the other hand, the indicative price offered by contracting firms at the beginning of the selling season is often very low in order to take account of a possible fall in the offer price later in the season. Prices for sunflower seed do not tend to fall during a season, but they are usually lower at the beginning of the season due to the scale of the supply. Thus, independent traders who flock to the villages and buy directly from farmers for cash intensify competition and provide an opportunity for farmers to maximise their sunflower revenue through side-selling. Farmers attribute this situation to the firms' low level of commitment to contractual obligations, such as collecting the harvested products frequently and supplying adequate inputs. According to the farmers, contracting firms lack the right attitude, preferring instead 'to reap more than they have sown'. As independent economic agents, farmers feel they have the freedom to choose whom they can sell their produce to once they have harvested.

However, the owners of the contracting firms claim that this situation is a result of the unfriendly business environment that currently prevails in the sub-sector. Contracting firms have not yet been able to influence the existing legal environment by, for instance, proposing some kind of zoned monopsony. The lack of a specific legislative framework to support contract farming is mentioned as the reason why trust cannot be high. Farmers give the same reason for firms' poor levels of commitment. This is a major reason for the lack of trust between the parties, and it is also the main source of contractual weaknesses, including side-selling.[6]

Concluding remarks

This study has shown that successful contract farming goes beyond the initial sales agreement between smallholder farmers and contracting SMEs. This examination of seven contractual arrangements involving different SMEs suggests that successful cases of contract farming should include a comprehensive package of additional elements. Furthermore, the findings emphasise the role and importance of various other actors in the value chain, such as government institutions, NGOs and business development organisations. For many firms, the share of products from contract suppliers through a coordinated supply chain is still very small, and most firms lack the financial resources to ensure adequate access to production support services for farmers. The weak facilitation of expected service delivery fails to foster trust between farmers and firms, and side-selling emerges as a stumbling block in ensuring sustainable contract farming.

The analysis in this chapter also indicates that the probability of a local small-scale sunflower oil-processing firm effectively influencing the value chain depends on whether the firm can successfully coordinate seed delivery services, thus ensuring the availability of sufficient quantities of quality varieties at affordable prices while also providing profitable product markets. As the recorded increase in sunflower seed production is a result of the application of improved seeds, availability of seed is of the essence. Hence, seed supplies need to be increased from their current low level as compared to potential demand. The implication for policy is that the capacity of the public- and private-sector institutions that are responsible for the mass production of QDS should be expanded. Both ASA and TOSCI need to be reformed and their capacity enhanced in order to scale up the supply of improved seeds from agricultural research institutes while also setting up suitable procedures for the production and certification of quality declared seeds from selected villages.

The findings also indicate that, if the contracting SMEs are able to facilitate smallholders' access to knowledge, skills and specific good crop-husbandry practices, the value chain will on the whole be strengthened through 'tighter' and better organised coordination. Large-scale firms with superior technology and links to the high-end market could also play a significant coordination role. Currently large-scale processing firms provide a market opportunity for seed cake from SMEs. With an installed capacity of more than 900,000 tons, which is much greater than the current level of vegetable seed production nationally, large-scale firms offer a useful conduit for SME products through further processing of the unrefined oil from the SMEs. The crude edible oil from SMEs, which is the final product of most small and medium-size firms, still contains impurities such as wax, iron sediment, gum and water (FAO, 2012), making it less suitable for human consumption. Hence, the product is classified by both the Tanzania Food and Drugs Authority (TFDA) and the Tanzanian Bureau of Standards (TBS) as low in quality. From an economic point of view, it would be a better option to utilise the existing capacity of the larger firms as opposed to UNIDO's current plans to build a joint double refining plant for SMEs. Since this may not occur spontaneously, initiatives along the lines of sub-contracting and exchanges between large- and small-scale firms need to be worked out, possibly with UNIDO's involvement, and based on international experience.

One straightforward policy implication is that the government and other stakeholders need to explore better options that will provide adequate financial resources to private SME investors. Commercial banks in Tanzania are risk-averse, and the cost of capital, especially for SMEs, is very high. In addition, it is difficult for many firms to exploit the available credit facilities due to a lack of sufficient collateral. It is thus important to develop a financial regulatory environment that is conducive to the specific needs of SMEs involved in contract farming to establish supplementary activities on top of their commercial transactions with smallholders. For example, the establishment of a specialised bank for contract farming in agriculture would allow both smallholders and firms to borrow the funds required to finance their agricultural activities. The policies which already exist, like the Small and Medium Enterprises Policy, need to be made more effective so as to be able to address emerging issues pertaining to contract farming. Hence, the development of a specific policy on the financing of contract farming systems managed by SMEs is called for to enhance market development in current market conditions for smallholder farmers.

Notes

1 Sampling of the SMEs was deliberately based on the criterion that a processor had operated contract farming at least once in the previous three years. Structured interviews were used to obtain answers on specific issues, while unstructured interviews were used to obtain the broader views of respondents on issues of interest to the functioning of the value chain. As the number of respondents was small, the interview guides were administered on a face-to-face basis. The interview lasted for about one hour, and individual respondents were interviewed in their homes, offices or selected places after an initial appointment. Respondents were interviewed once, and their responses were recorded directly into a voice recorder.

2 A price charged at the seed research centre by the ASA.
3 A kilogram of QDS cost 1,500 Tsh per kg. A bag, usually 70 kg, earns a farmer 1,500 × 70 = 105,000 Tsh, while 1 kg of sunflower seed produced at harvest time is 500 Tsh; hence a 70-kg bag earns a farmer only 500 × 70 = 35,000 Tsh (computed using field data).
4 Bahi District was formerly a part of what is now Dodoma town before new demarcation in 2007. It is thus a nearby rural district.
5 There is usually a price difference between purchases made in villages and those made in urban areas. It costs about 32,000 Tsh per bag of sunflower seeds during the peak harvest season in villages, while the same bag costs 35,000 Tsh in town. Buyers keep the village price a little lower to compensate for the costs of transport and accommodation they incur while collecting seeds in the villages, a factor known to all parties.
6 Issues of trust in the sunflower seed value chain are discussed further in Chapter 10, this volume.

References

ARI Ilonga (2008). *Survey on the Status of Sunflower Production in Selected Regions of Tanzania.* Kilosa, Tanzania: Agricultural Research Institute.

Bogetoft, P. and Olesen, H. B. (2004). *Design of Production Contracts: Lessons from Theory and Agriculture.* Copenhagen: Copenhagen Business School Press.

FAO (2011). *The State of Food and Agriculture: Women in Agriculture – Closing the Gender Gap for Development.* Rome: FAO.

FAO (2012). *Guiding Principles for Responsible Contract Farming Operations.* Rome: Office of Knowledge Exchange, Research and Extension, FAO.

Gereffi, G. (1999). International Trade and Industrial Upgrading in the Apparel Commodity Chain. *Journal of International Economics, 48*, 37–70.

Gereffi, G., Humphrey, J. and Sturgeon, T. (2005). The Governance of Global Value Chains. *Review of International Political Economy, 48*(1), 78–104.

Hawkes, C. and Ruel, M. T. (2011). *Value Chain for Nutrition.* Paper prepared for the IFPRI 2020 International Conference: Leveraging Agriculture for Improving Nutrition and Health, New Delhi, India.

MMML (2010). Benefits of Edible Oil Seeds Locally Planted and Produced. Presentation by Mount Meru Miller Limited on behalf of Local Edible Oil Seed Industry in Tanzania at a meeting with the Tanzania Development Partners Group. Online: www.tzdpg.or.tz/fileadmin/documents/dpg_internal/dpg_working_groups_clusters/cluster_1/psdtrade/Documents/Sunflower_presentation_Mount_Meru_Millers_ltd.pdf. Accessed 25 September 2013.

Mpeta, D. F. (2015). Effects of Contract Farming on Production and Income of Sunflower Farmers in Kongwa District in Central Agricultural Zone of Tanzania. Unpublished PhD Thesis. Mzumbe University, Tanzania.

Porter, M. (1985). *Competitive Advantage.* New York: Free Press.

Raynolds, L. T. (2002). Wages for Wives: Renegotiating Gender and Production Relations in Contract Farming in the Dominican Republic. *World Development*, 30(5), 783–798.

RLDC (2008). *Sunflower Sector: Market Development Strategy.* Dodoma, Tanzania: Rural Livelihood Development Company.

RLDC (2009). *Making Market Work for Producers.* Dodoma, Tanzania: Rural Livelihood Development Company.

RLDC (2012). *Success Stories, Issue No. 1, 16*. Dodoma, Tanzania: Rural Livelihood Development Company.

Singh, S. (2002). Contracting out Solution: Political Economy of Contract Farming in the Indian Punjab. *World Development*, 30(9), 1621–1638.

UNIDO (2009). *Value Chain Diagnostics for Industrial Development: Building Blocks for a Holistic and Rapid Analytical Tool*. Working Paper. Vienna: United Nations Industrial Development Organization (UNIDO).

UNIDO (2011). *Pro-poor Value Chain Development: 25 Guiding Questions for Designing and Implementing Agroindustry Projects*. Practitioner's Guide. Vienna: United Nations Industrial Development Organization.

URT (2002). *Small and Medium Enterprise Development Policy*. Dar es Salaam: Ministry of Industry and Trade.

URT (2003a). *The Tanzania Food, Drugs and Cosmetics Act*. Dar es Salaam: Government Printer.

URT (2003b). *National Trade Policy: Trade Policy for a Competitive Economy and Export-Led Growth*. Dar es Salaam: Ministry of Industry and Trade.

URT (2015). *Speech of the Minister of Agriculture, Food and Cooperative Christopher K. Chiza (MP) on the Budget for the Year 2015/16*. Dar es Salaam: Ministry of Agriculture, Food and Cooperatives.

URT (2016). *Sunflower Sector Development Strategy 2016–2020*. Dar es Salaam: United Republic of Tanzania.

URT/FAO (2008). *Agricultural Sector Reforms in Tanzania: Perspective from Within*. Dar es Salaam: Ministry of Agriculture Food Security and Cooperatives, in collaboration with FAO, Rome.

WEF (2010). *Realizing a New Vision for Agriculture: A Road Map for Stakeholders*. World Economic Forum. Online: www3.weforum.org/docs/WEF_IP_NVA_Roadmap_Report.pdf. Accessed 5 August 2013.

6 Contract farming and upgrading possibilities for smallholder sugarcane-growers in Tanzania

Thobias E. Nsindagi and Jennifer K. Sesabo

Introduction

The issue of upgrading smallholders' production processes has increased in importance following what seems to have been the global adoption of privatisation and market liberalisation policies, which have created global competition among producers of primary commodities. Regrettably, smallholders in developing countries can hardly compete globally due, among other things, to poor access to productive resources. Tanzania is no exception, as most of its smallholders lack the capacity to compete with food imports from more cost-effective producers in other countries. This is also the case in the sugar sector, where the ineptitude of smallholder sugarcane-growers is undoubtedly associated with a lack of thoughtful and supportive initiatives geared towards accessing services with the potential for improving production performance. The literature suggests that contract farming arrangements can facilitate the availability of important services to farmers. Similarly, as contract farming integrates smallholder farmers into modern markets, it may facilitate production by providing services and productive inputs such as seeds, fertiliser, pesticides, on-farm training, technical services, access to credit, extension services, land preparation and hauling of produce (Singh, 2005; Setboonsarng *et al.*, 2008; Prowse, 2012).

The government of Tanzania also claims that contract farming promotes public–private partnerships and accelerates technology transfers, capital inflows and the development of product markets (URT, 2013). The aim of this chapter is to describe how sugarcane contract farming in Tanzania influences the performance of smallholders. The chapter is organised as follows. The next section presents the history and current status of sugarcane farming in Tanzania. This is followed by the conceptual framework in which definitions of the key concepts of capability acquisition and incentives are outlined. Subsequently, a description of the methodology used is followed by an outline of the findings of the study, a discussion of the barriers to process upgrading and a conclusion with policy recommendations.

The history and current status of sugarcane farming in Tanzania

The sugar sector in Tanzania started in the early 1920s when the Tanganyika Plantation Company (TPC) established a factory in Kilimanjaro. This was followed several decades later by the launch of two companies in Morogoro, namely the Kilombero Sugar Company in 1962 and the Mtibwa Sugar Company in 1963. In 1982 Kagera Sugar Limited (KSL) was substantially expanded and reopened by the government after the Tanzania–Uganda war ended in 1978. From the late 1960s, these companies were 100 per cent owned by the government and managed by the National Food Corporation (NAFCO), which later, in 1974, was replaced by the Sugar Development Corporation (SUDECO). Management of sugarcane farming by smallholders was carried out within the framework of out-grower schemes, with the government being responsible for the provision of inputs and the training of out-growers through various state agencies such as the Kilombero Sugar Institute, the Kibaha Sugarcane Research Institute and the Ministry of Agriculture through its field extension department.

In 1998 Kilombero was privatised and taken over by Illovo Sugar Limited (ISL, a South African Company) and ED&F Man Holdings Limited (a company based in the United Kingdom) with equity shares of 55 per cent and 20 per cent respectively; the remaining 25 per cent equity share was retained by the government. The Kilombero Sugar Company Limited (KSCL) is now the largest sugar-processing factory in the country, and it still manages an out-grower scheme with small-, medium- and large-scale farmers. A local private company, Tanzania Sugar Industries Limited (TSIL), fully acquired the Mtibwa Sugar Company in 1999 and Kagera Sugar Limited in 2001. In 2000, TPC sold a 75 per cent share to the Sukari Investment Company Limited, a company registered in Mauritius and owned by Deep-River Beau Champ Limited and Quartier Français; the government retained a 25 per cent share. TPC and KSL are not included in this study because the former is not involved in out-grower schemes and the latter only started an out-grower scheme recently. Hence, both the out-grower schemes examined in this chapter are located in Morogoro Region.

After privatisation, various services offered to smallholder cane-growers gradually ceased as the government started to withdraw from direct involvement in productive activities. Following the withdrawal of the government from providing assistance to farmers, the head of the out-growers' association, the Tanzania Sugarcane Growers Association (TASGA), pushed to get contracts that could help farmers secure better prices and loans from banks using their sugarcane fields as collateral. In 1999, one year after the privatisation of KSCL, TASGA drafted the first contract, but Illovo rejected it and returned it with a counter-draft in 2000. By 2006, after years of sending revised drafts back and forth, Illovo returned the forty-second revised draft to TASGA with a 'take it or leave it' ultimatum. Finally, by late June 2006, the out-growers' associations and the company signed the first contract in Kilombero. Subsequently, in 2009, the first contract was signed in Mtibwa (Turiani division).

From then on, contract farming in Morogoro's sugar sector has been governed by the contents of the Cane Supply Agreement between the out-growers' associations and the two processing companies ('millers'). The clauses in the Cane Supply Agreement include the supply of cane, the cane delivery programme and the miller's obligation to crush cane that meets the specified quality. Other clauses cover the determination of quality (measurement of sucrose content or so-called rendement), purchase price, payment modalities, cane varieties and dispute settlement procedures.

With the introduction of contract farming and privatisation in general, factories were rehabilitated and their capacities improved. After the privatisation of the sugar sector, trade in sugar was also liberalised in an effort to cater for the increasing demand for it (Sugar Board of Tanzania [SBT], 2014). Generally, sugarcane production has showed a growing trend for the past 50 years (1961/62–2009/10) (FAO, 2011). However, production trends differ when production is disaggregated between the out-growers and the estates owned by the milling companies. For example, out-grower production in Kilombero and Turiani divisions for 1996/97–2012/13 provides an ambiguous picture regarding the performance of out-growers. The trends indicate that there has been steady growth in sugarcane production in Kilombero up to a fairly constant level of about 500,000 tons, although marred by serious setbacks in the 2006/07 and 2009/10 seasons caused by severe droughts (SBT, 2014). However, the trend in Turiani shows that in 1997/98 smallholder sugarcane production in Turiani was exceeded by production in Kilombero. Moreover, from 2003/04, Turiani's out-grower sugarcane production lagged far behind that of Kilombero, stagnating at a much lower level of about 200,000 tons and experiencing a continual decline from 2008/09 (see Figure 6.1). The fact that production in one of the out-grower schemes seems to be stagnating while production in the other has clearly declined from about the time contract farming was introduced poses some interesting questions. Do contract farming arrangements have any influence on these observed trends? How did contract farming arrangements affect smallholder performance? This chapter investigates the extent to which the existing contractual arrangements in the sugar sector in Tanzania facilitate or obstruct the upgrading of out-growers' production and points to the policy measures that are needed to increase production in this sub-sector.

Contract farming and upgrading: a capability perspective

A contract is an important institution in which production, distribution and marketing arrangements can be coordinated between the actors in a value chain. Da Silva (2005) defines contract farming as a general term referring to a particular form of supply chain governance adopted by firms to secure

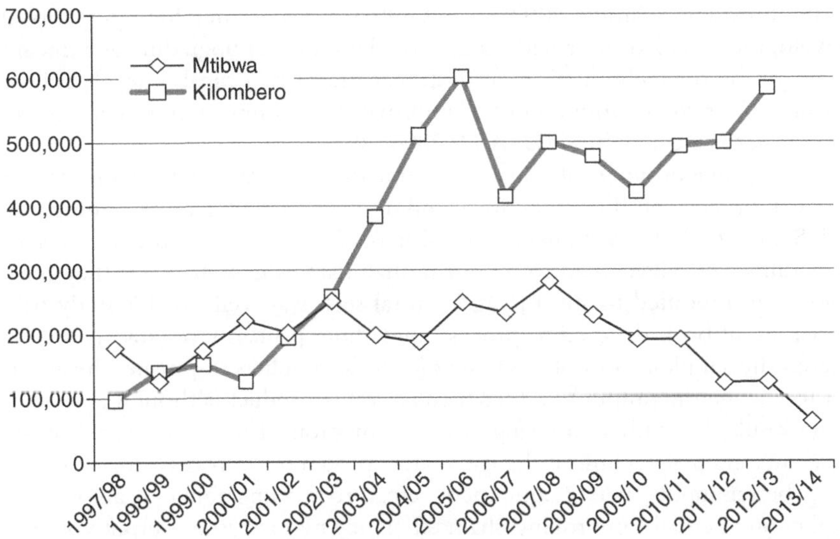

Figure 6.1 Sugarcane production trend for outgrowers in Kilombero and Mtibwa (Turiani).

Sources: Kilombero (2014); Mtibwa (2014).

access to agricultural products and raw materials that meet the desired specifications of quality, quantity, location and timing. Prowse (2012: 10) defines contract farming as:

> a contractual arrangement between a farmer and a firm, whether oral or written, which provides resources and/or specifies one or more conditions of production, in addition to one or more marketing conditions, for an agricultural product, which is non-transferable.

Prowse's definition emphasises the provision of resources and enhancement of market access for produced outputs. This implies that the adoption of contract farming arrangements based on this definition will benefit resource- and market-constrained smallholders, although it might also entail higher costs for the firms engaged in the contract.

As a concept, upgrading is conceptualised and used in several different contexts. Many practitioners of the global value chain (GVC) approach concur with Gereffi (1999) and Humphrey and Schmitz (2002) that upgrading is a result of the interaction between so-called lead firms and local producers, which consequently add new and improved capacity to local producers, so they are able to reap higher benefits from participation in the value chain. The capacity of local producers is enhanced through learning by doing and through the lead firm allocating new tasks to local suppliers (Humphrey and Schmitz, 2002). The United States Agency for International Development (USAID) provides a generic definition of upgrading as a 'means for acquiring the technological, institutional and market capabilities that allow resource poor rural communities to improve their competitiveness and move into higher value activities' (USAID, 2006: 2).

Usually practitioners of the GVC approach (see Chapter 1) distinguish between process, product, functional and inter-sectoral upgrading (Humphrey and Schmitz, 2002; Kaplinsky and Morris, 2008). Initially these forms of upgrading were developed to serve the manufacturing sector, but they have since been modified to suit the agricultural sector as well. In this study particular attention is devoted to process upgrading, primarily because it is considered the simplest form of upgrading for local suppliers to pursue: the aim is not to change the output to a higher (unit) value product (although that often is the result) but rather to change the organisation of production and introduce incremental technical change in the production process, for instance, using better inputs, better equipment, etc. Process upgrading in an agricultural context entails 'improving chain efficiency by increasing output volumes or reducing cost for a unit of output. For instance, improving agronomy to enhance yields which may involve the use of improved technologies, planting materials, or supply of inputs' (Mitchell *et al.*, 2009: 33).

Contract farming and upgrading can be linked by incorporating two broad concepts, namely the capability for upgrading and the incentive for upgrading (Humphrey, 2004; see also Young and Hobbs, 2002). While transaction-cost theory (e.g. Kumar, 2006; Singh and Asokan, 2005) emphasises minimising the costs of the economic exchange to preserve the business relationship, the capability approach emphasises the existence of mutual competencies among actors in the contractual relationship (Prowse, 2012). It is assumed that it is almost impossible to maintain a relationship if the respective actors have capabilities that cannot be aligned. Young and Hobbs (2002) argue that the internal management of contract farming arrangements requires individual or team competencies, skills and tacit knowledge to help maintain the organisation of the arrangement. Likewise, Prowse (2012: 35) asserts that 'the ability to create and sustain contract farming operations relies to large extent on the skills and experience of staff and the ability of the organisation to maximise these'. Capability in this context means the ability to perform the assigned tasks of the contractual relationship. For example, farmers should have the ability to produce according to the technical specifications. Capability can also be considered to mean the ability of individuals to learn and apply the knowledge and skills they have acquired to undertake a particular activity. Contractual relationships that fail to facilitate sharing, learning and the synthesis of endowed capabilities will struggle to exist, and the prospects of (process) upgrading will be severely hampered.

Upgrading capability can be realised by means of learning processes that occur either through learning by doing or learning first and doing later. This is the main idea behind the traditional learning-by-exporting paradigm, which emphasises the superiority of exporting for learning and industrial upgrading. The learning parameters include training, extension services and technical assistance to farmers and the adoption of best farming practices. These parameters are important for building the necessary capacity for smallholders' performance and upgrading (Bolwig *et al.*, 2009). Studies have indicated that the parameters of a successful learning process (i.e. training, extension services and adoption of the best farming practices) are terms often used in contract farming packages (Prowse, 2012). Empirical evidence from different crops also indicates the importance of training, extension services and best farming practices in enhancing the upgrading of smallholder farmers involved in contract farming arrangements in Indonesia, Latin America, Zambia, Uganda, India and Kenya (see, for example, Bolwig *et al.*, 2009; Emongor *et al.*, 2004; Grosh, 1994; Humphrey, 2004; Mitchell *et al.*, 2009; USAID, 2006).

Learning processes for building up smallholder capacity are often constrained by a lack of physical and financial resources, potentially undermining even modest technological improvements. Physical capital includes farming equipment and agricultural inputs such as fertiliser and seeds, while financial resources include access to credit or loans. According to Barrett *et al.* (2012), a lack of agricultural inputs and limited access to credit hinders the production capacities of smallholder farmers in developing countries. Access to capital and financial resources is therefore an indicator of the potential acquisition of capabilities (Humphrey, 2004; Mitchell *et al.*, 2009). As already mentioned, contract farming arrangements have been considered a solution to the problems of input and credit access. Several studies acknowledge the importance of resource-provisioning contracts under which agricultural inputs are provided inexpensively or on credit and in kind in the form of short-term loans (Jaffee *et al.*, 2011; Prowse, 2012).

However, in order for smallholders to acquire the requisite capabilities, it is necessary to examine the existence of a set of equally important parameters for process upgrading, namely what are here called incentive parameters. These parameters are to be conceptualised as *sine qua non* conditions: they are necessary incentives for smallholders to start venturing into any kind of capability acquisition and learning process. They include output price (market access), availability of supply, infrastructure and logistics, payment mechanisms and institutional issues (Humphrey, 2006; USAID, 2006). For instance, several studies have documented the crucial importance of output prices for the performance and process upgrading of smallholders worldwide. Thus, coffee farmers in Uganda upgraded their processes to produce organically, and net income increased as price premiums were assured (Bolwig *et al.*, 2009). Other examples include conventional and organic cotton smallholder production in India (Eyhorn *et al.*, 2005) and fishermen on the shores of Lake Victoria in Tanzania (Kadigi *et al.*, 2007).

Methodology

Process-upgrading possibilities for smallholders in the two out-grower schemes were assessed by evaluating the two concepts discussed above, that is, the capability for upgrading and the incentives for upgrading. The key concepts are elaborated on the basis of the framework used by USAID (2006). According to the framework, the parameters that constitute capability for upgrading include access to inputs, credit, training, extension services and production information. Incentive parameters include the price of sugarcane, determination of rendement (i.e. as a proxy for sucrose content), payment mechanisms and contractual enforcement. Simple 'binomial' perceptions from different categories of respondent in the two schemes were used to assess the key parameters. The respondents were interviewed between May and June 2013 using open-ended, semi-structured interview guides. The interviews were conducted after a quantitative survey was conducted from July to August 2012. The number of interviewed respondents, divided into different categories, is shown in Table 6.1.

All the producers lived in four villages, two in each scheme (Kitete and Kidatu in Kilombero, and Kilimanjaro and Lungo in Mtibwa/Turiani). The producers are smallholders selected on the basis of farming and harvesting zones; smallholders living in the same village may have fields that are widely separate from one another in the sugarcane catchment area. There are four zones categorised according to the distance from the factory's location. Zone I covers sugarcane fields located 0–5 km away from the factory gate, zone II at 6–10 km, zone III at 11–15 km and zone IV at 20 km and above. Two farmers from each zone who had been engaged in sugarcane cultivation for the previous 12 months were selected for interviews with the assistance of farmers' group leaders and subsequent 'snowballing'.

Farmers' group leaders were also included in the study because the organisation of contract farming is based on the farmers' groups that are found in all communities. Of the nine farmers' group leaders interviewed, three were from communities in Turiani (Kilimanjaro – TUCO, Kilimanjaro – MOA and Lungo – TUCO) and six from Kilombero (Kitete, Upendo, Harakati, Mkanyageni, Mwangaza and Sunguburu).

Finally, the facilitators are farmers' associations with responsibilities like assisting farmers to secure loans from financial institutions (including SACCOS) and

Table 6.1 Overview of respondents

Unit name	*Category*	*Number*	*Location*
Producers	Smallholder farmers	8 × 4 villages	Two in each scheme
Groups	Smallholder farmer groups	9	Kilombero (6); Turiani (3)
Facilitators	Farmer associations	14	Kilombero (12); Turiani (2)

Source: authors' fieldwork (2012).

offering extension services, farming and harvesting services, advocacy and lobbying services. The associations represent out-growers in the drawing up and negotiation of contract terms. A total of 14 associations were selected based on location and size in terms of the number of registered and active members (see Appendix 6.1).

Perceptions of incentive parameters for process upgrading

This section reports the results from the interviews with the three categories of actor (i.e. farmers, farmers' groups and associations) concerning their perceptions of the selected incentive parameters. The results are outlined in Table 6.2 for Kilombero and Table 6.3 for Turiani.

Pricing arrangements

One finding is that 62.5 per cent of farmers, 33.3 per cent of farmers' groups and 50 per cent of associations in Kilombero considered pricing arrangements unfair. In Turiani 100 per cent of all respondents (farmers, farmers' groups and associations) were of the same opinion. All respondents (farmers, farmers' groups and associations) felt that this was because they had no negotiating power over seasonal prices, so they considered themselves to be price-takers.

Table 6.2 Perception towards incentive parameters for process upgrading in Kilombero

Category	*Perception*	*Percentages*		
		Farmer N = 16	*Groups N = 6*	*Assoc. N = 12*
Pricing arrangement	Fair	37.5	66.7	50
	Unfair	62.5	33.3	50
Sugarcane supply	Increasing	12.5	16.6	8.3
	Decreasing	62.5	66.7	75
	Fluctuating	25	16.6	16.7
Rendement determination	Fair	25	16.7	
	Unfair	75	83.3	100
Payment duration	Timely	31.2	33.3	8.4
	Delayed	68.8	66.7	91.6
Availability and efficiency of harvesting equipment	Yes			16.7
	No	100	100	83.3
Enforcement effectiveness	Ineffective			16.7
	Absent	100	100	83.4

Source: authors' fieldwork (2012).

Table 6.3 Perception towards incentive parameters for process upgrading in Turiani

Category	*Perception*	*Percentages*		
		Farmers N = 16	*Groups N = 3*	*Assoc. N = 2*
Incentive				
Pricing arrangement	Fair			
	Unfair	100	100	100
Quantity of produce	Increasing			
	Decreasing	100	100	100
	Fluctuating			
Rendement determination	Fair			
	Unfair	100	100	
Payment duration	Timely			
	Delayed	60	100	100
	Extreme delay	40		
Availability and efficiency of harvesting equipment	Yes			
	No	100	100	100
Enforcement effectiveness	Effective			
	Ineffective			
	Absent	100	100	100

Source: authors' fieldwork (2012).

Sugarcane supply

With regard to trends in sugarcane supply, 62.5 per cent of farmers, 66.7 per cent of farmers' groups and 75 per cent of associations in Kilombero asserted that the quantity of produce had fallen over time, and in Turiani too all respondents (i.e. 100 per cent of farmers, farmers' groups, associations and the miller) asserted that the quantities of sugarcane being supplied by smallholders were declining tremendously. The reasons for the decline in supplies, according to farmers in Kilombero, were the low rendement calculated for their produce, bad harvesting management, including non-adherence to the harvesting schedule, and cheating in weighing cane. The associations also pointed out that the small daily quotas for crushing allocated to them by the processing company was one of the most significant factors constraining greater participation in sugarcane production. In Turiani, farmers, farmers' groups and associations gave similar reasons, namely that they were adversely affected by delayed payments, bad harvest management, low cane prices, too many deductions and too much power on the part of the miller. A study by Kamuzora (2010) finds analogous results whereby some farmers could harvest their sugarcane every year, even though their names were not even listed, while those farmers whose names were listed in the harvesting schedule were passed over, giving rise to allegations of bribery. Similarly, REPOA (2012) observes comparable findings that harvest operations were considered inefficient and unfair, and concluded that, to ensure a fair and efficient harvest rotation, all the cane areas in Kilombero should be divided into zones that will all be overseen by just one association acting as a trust.

Determination of rendement

Perceptions regarding the determination of rendement in Kilombero indicate that 100 per cent of the associations, 75 per cent of farmers and 83.3 per cent of farmers' groups that were interviewed felt that rendement was determined unfairly. In Turiani, all respondents except the miller said the same. It was considered unfair because the process was not transparent and violated the agreed provisions. It was agreed that rendement should be tested before the sugarcane is harvested instead of the current practice of it being determined after harvesting and crushing had taken place.

Payment duration

Pertaining to payment modalities, 68.8 per cent of farmers, 66.7 per cent of farmers' groups and 91 per cent of associations in Kilombero stated that payments were delayed, while 31.2 per cent of farmers, 33.3 per cent of farmers' groups and 8.4 per cent of associations felt that payments were made in a timely fashion. In Turiani, none of the respondents agreed with the latter statement. Instead, 60 per cent of farmers declared that payments were delayed, while 40 per cent said that payments were extremely delayed. Farmers' groups, associations and the miller claimed that payments were delayed. Of those in Turiani who argued that payments were delayed, 60 per cent stated that payment periods were 3–5 months, 40 per cent that they were extremely late at 5–6 months. In Kilombero 60 per cent of all respondents stated that payments were made at 1–3 months and 40 per cent that they were made within a month. Delays in paying out-growers for their sugarcane deliveries to the mill were also reported by Nkonya and Barrero-Hurle (2012) and Nazir *et al.* (2013).

Availability and efficiency of harvesting equipment

Concerning infrastructure and logistics – that is, the availability and efficacy of harvesting, loading and transportation equipment – findings revealed that 100 per cent of farmers and farmers' groups in Kilombero considered that harvesting equipment was insufficient to meet demand, while 16.7 per cent of associations claimed that equipment belonging to the associations was available and efficient in Kilombero. In Turiani no respondent was of this opinion (Table 6.3). In this case, farmers used to engage sub-contractors to do the cutting, loading and hauling of their sugarcane, although these sub-contractors were not competent enough to perform all the functions assigned to them. Similar observations were documented in the Tanzania Sugar Industry Development Plan and Strategy (2011/12–2015/16). According to this document, it is estimated that 10 per cent of the crop is left unharvested in the fields each year due to poor infrastructure and logistics.

Enforcement effectiveness

All respondents (i.e. farmers, farmers' groups and associations) in both locations claimed that enforcement of the contractual arrangements was lacking.

Perceptions of the capability parameters for process upgrading

This section reports on the perceptions of respondents regarding the availability of the capability parameters for process upgrading (i.e. training received on farm management, access to extension services, timely supply of fertiliser, production and marketing information, adequate application of fertiliser, knowledge of rendement and understanding of price determinations). The results are outlined in Table 6.4 for Kilombero and Table 6.5 for Turiani (as above).

Table 6.4 Perception towards capability parameters for process upgrading in Kilombero

Category	*Perception*	*Percentages*		
		Farmer N = 16	*Groups N = 6*	*Assoc. N = 12**
Availability of training	Yes	56.2	16.7	58.3
	No	43.8	83.3	41.7
Access to extension services	Yes	18.7	16.7	25
	No	81.3	83.3	75
Timely supply of fertiliser	Yes			33.3
	No	100	100	66.7
Production and marketing information	Yes			16.7
	No	100	100	83.3
Fertiliser applied adequately	Yes		20	16.7
	No	100	80	
Access to credit	Yes	68.7	66.7	83.3
	No	31.3	33.3	16.7
Knowledge of measurement of rendement	Yes			16.7
	No	100	100	83.3
Understanding price calculation	Yes	25	33.3	8.3
	No	77	66.7	91.7

Source: authors' fieldwork (2012).

Table 6.5 Perception towards capability parameters for process upgrading in Turiani

Category	*Perception*	*Percentages*		
		Farmers N=16	*Groups N=3*	*Assoc. N=2*
Availability of training	Yes	18.75	66.7	
	No	81.25	33.3	100
Access to extension services	Yes	6.2		
	No	93.8	100	100
Timely input supply	Yes			
	No	100	100	100
Production and marketing information	Yes			
	No	100	100	100
Input fertiliser applied adequately	Yes			
	No	100	100	100
Access to credit	Yes	31.2		100
	No	68.8	100	
Knowledge of measurement of rendement	Yes			100
	No	100	100	
Understanding price calculation	Yes	12.5	33.3	
	No	87.5	66.7	100

Source: authors' fieldwork (2012).

Availability of training

Responses regarding the availability of training in Table 6.4 reveal that 56.2 per cent of farmers, 16.7 per cent of farmers' groups and 58.3 per cent of the associations interviewed in Kilombero acknowledged that they received training. In Turiani, the corresponding figures were 18.75 per cent of farmers and 66.7 per cent of farmers' groups. They claimed that the company, SUDECO, the Sugar Research Institute based in Kibaha and Daipesa (NGO), conducted training just once, and that a long time ago, to individuals who were either group leaders or had leadership positions in the community, especially in Turiani. Conversely, 100 per cent of associations and more than 80 per cent of farmers in Turiani had never received any training at all. Respondents stated that there was no provision in the agreement regarding the training of smallholder cane-growers.

Access to extension services

More than 80 per cent of farmers and farmers' groups alleged that it was difficult to access extension services in Kilombero, while in Turiani more than 90 per cent of farmers and 100 per cent of farmers' groups and associations stated the same. Access to extension services did not form part of the agreement between millers and sugarcane out-growers.

Timely supply and application of fertiliser

It was noted that 100 per cent of farmers and farmers' groups in both areas claimed that they could neither receive fertiliser in a timely fashion nor apply enough fertiliser. This finding corroborates the survey database of 200 farmers surveyed in Kilombero, which shows that fertiliser usage was only 58 per cent of the required amount per hectare, while the 200 Turiani farmers used only 18.7 per cent of the required amount. Despite the importance of fertiliser in increasing yields per hectare, this is not included in the agreement. The respondents testified that the companies monitor smallholder sugarcane in an 'eyes on' and 'hands off' manner.

Production and marketing information

Concerning regular access to information about sugarcane production, the results from both locations show that 100 per cent of farmers and farmers' groups were not provided with information on sugarcane production. In principle such information could be obtained from agricultural extension officers or any other relevant authorities endorsed in the contract agreement. Unfortunately, however, there is no provision guaranteeing the distribution of production information to smallholder cane-growers in the study areas.

Access to credit

Regarding access to credit, 68.7 per cent of farmers, 66.7 per cent of farmers' groups and 83.3 per cent of associations agreed that it was easy to access credit in Kilombero, while in Turiani the figure was only 31 per cent of farmers. Neither farmers' groups nor associations felt that access to credit was easy in Turiani.

Measuring rendement and price calculations

Concerning knowledge regarding the measuring of rendement and price calculations, 100 per cent of farmers and farmers' groups in both areas claimed to be unaware of how rendement was determined. In addition, more than 75 per cent of farmers and 65 per cent of farmers' groups in Kilombero and Turiani claimed not to understand how the price of sugarcane was calculated.

Is process upgrading happening?

The objective of this chapter has been to examine the potential of process upgrading for out-grower production in the Tanzanian sugar value chain. The analysis was guided by the following questions: do contract farming arrangements have any influence on out-growers' performance? And how do contract farming arrangements affect potential of out-growers to upgrade their production processes?

Assessment of the factors affecting process upgrading was based on the capability and incentive parameters established theoretically. Based on the results presented in the previous section, the range of incentives for smallholders to acquire capabilities facilitating the greater adoption of process upgrading was found to be limited. None of the parameters that have been recognised in the literature as sources of the learning and of the acquisition of capabilities were fully realised in the contractual arrangements. The results emphasised that smallholder performance had deteriorated over time, especially in Turiani. The most compelling reasons mentioned were a lack of incentives stemming from unfair sugarcane prices, delayed payments, unfair determination of rendement and poor infrastructure, which resulted in a lack of capability acquisition among sugarcane-growers.

With regard to sugarcane prices under contract farming arrangements, these are in principle negotiated and agreed among the actors before the buying and selling season. However, the price offered under the contract farming arrangements in both study areas is not a result of mutual agreement between out-growers and millers. The miller usually sets the price, and the out-growers become price-takers. Out-growers are given a complex contract containing technical information on price determination to study and decide upon under great time pressure. Normally this is distributed during an annual general meeting, often with little or no detailed explanation on the part of the company. Unfortunately, the majority of cane-growers lack legal or other professional skills and have no access to such services. Even the leaders of the farmers' association could not explain the relevant mechanism and parameters with regard to price calculation. This has demotivated farmers from improving their production processes because they think that prices are unfairly determined. They are forced to sell their sugarcane at the price dictated to them due to the monopsony situation created by the country's regulatory framework.

Many farmers are used to fixed prices and do not realise that prices in the market fluctuate. In the negotiations on prices, farmers seem to think that prices should always reflect 'their efforts'. The poor understanding of how prices for sugarcane and rendement are determined indicates a problem with the flow of information within the value chain. This finding endorses Prowse's verdict (2012) that information and communication blockages exist in contract farming arrangements. In order to foster competitiveness, producer organisations should facilitate knowledge-sharing between members to integrate them into the value chain.

Similarly, late payments have been the norm, although they are contrary to the agreement, which requires farmers to be paid every two weeks from the last month of cane delivery. However, this is clearly not the case, and payment is often delayed by up to three to five months, or even more, especially in Turiani. This failure to adhere to the agreement has affected farmers' other economic activities, such as timely weeding, fertilising and pesticides applications, replanting and loan repayments. Generally, non-adherence to the payment modalities agreed in the contracts has been a major problem for farmers.

The fact that smallholder sugarcane-growers lack capability acquisition for their production process is visible in the low quality of sugarcane they produce. Respondents claimed that the processing companies assessed the rendement of their sugarcane as low, which is a sign of poor quality. Rendement is considered low if it is below 10 per cent, and it is very difficult for smallholders to achieve a rendement above this level.

Low rendement, which leads to low prices coupled with high production costs, prevents most out-growers from adopting the best production practices. For instance, a continuous practice among growers has been to harvest unmanaged sugarcane fields year after year without systematic replanting due to low net returns. Similar findings were obtained by Matango (2006), whose study reveals that 'farmers' cane fields are not weeded, costs rise as weeds intensify and canes produce little tonnage, [and] low sucrose without nutrients, and finally fetch little money to farmers' (p. 11). In a situation in which cane delivered for purchase receives low rendement, it becomes difficult to establish whether the reason is poor product quality caused by bad cane husbandry or sloppy harvest management as contractors and associations do not deliver on time. All in all, any act compromising product quality represents a violation of clause 5.2.2 of the agreement between out-growers and processors, which requires that all cane delivered to the mill must be in a good, clean and fresh condition substantially free of trash, tops, dirt and roots, suitable in all respects, and ready for milling and processing (MOA Cane Supply Agreement, 2011: 7).

Harvesting poor-quality cane and assigning it a low rendement actually offends against the harvesting protocol, which requires that sugarcane should be tested in the laboratory to confirm both crop-maturity and crop-quality before burning and cutting are carried out. Based on the 2009/11 agreement between out-growers and MSEL in Turiani, sugarcane should be accepted for delivery to the mill if it has a first expression juice (FEJ) purity of not less than 80 per cent tested before harvest. Yet, in practice, the out-growers' sugarcane sampling procedures are conducted after all harvesting operations such as burning, grabbing and hauling have already been done. Upon arrival at the mill-gate, the following sugarcane sampling procedure is carried out:

- i Cane is weighed at the weighbridge, and the following data are compiled and fed into the computer: name, out-grower number and location where the cane was harvested.
- ii At least five cane trucks (totalling approximately 50 tons of cane) passing through the weighbridge from the same location and date are given one and the same cane secret code number.
- iii The cane secret code number tag is given to the cane yard sample boy or girl for follow-up. The 50-ton pile of cane is fed on to the feed table, and the sample boy or girl follows the cane passing through the shredder to the first mill, where the juice sample is collected.

iv After collecting the juice from the first mill, the tag is submerged in the sample container, and preservative is added from the first mill to avoid juice deterioration.
v All the cane delivered from different locations not weighing at least 40–50 tons is assessed at an average rendement for all the out-growers on the specific day and date.
vi Cane weight notes are made available to the cane-grower and cane-transporter.
vii Rendement reports are made available to the farmers' association after calculations.

Such practices violate the agreement and expose the out-growers to marketing risks because the rendement assessment is not clear to out-growers, meaning that they are not sure of their sales revenues until they receive the final notification from the miller.

Apart from problems related to rendement determination, respondents also frequently cited harvesting operations across cases. For example, bad harvesting management was cited by 18.7 per cent of farmers in Kilombero and by 25 per cent of farmers and 33.3 per cent of farmers' group leaders in Turiani. Harvesting operations, which include cane cutting, loading and hauling, are often considered jointly due to the fact that sugarcane must be transported to the processing facilities within hours of harvesting to avoid spoiling. Normally, farmers assume the costs of cutting, grabbing and hauling, but they have no real control over the harvesting schedules.

Farmers also reported harvesting delays of up to one year or more for the mature cane due to non-adherence to the harvesting schedule. Some farmers reported that their fields were not harvested in consecutive years, while others had their fields harvested every year. According to a respondent in Kilombero, 'favouritism in harvesting is an immense problem, and something has to be done to rescue the organisation of harvesting process in the area'. Non-adherence to the harvesting schedule has caused smallholders increasing frustration for years, one indication of which is the frequent decision to circumvent the rules and burn the fields prematurely, causing what are called 'fire accidents' in order to ensure that crops are harvested. Fire accidents (in this sense) are unplanned burning practices that may cause severe problems in coordinating the overall harvest, as the total volume of sugarcane may exceed the factory's capacity.

Regarding institutional issues, state regulatory frameworks have created a monopsony situation in which out-growers cannot choose firms with which they want to do business, as the law governing the sugar sector does not allow the establishment of new milling factories in an area, normally within a radius of 40 km from the existing mill. In addition, the lack of effective enforcement mechanisms in the contractual agreements has been a source of problems to farmers in improving their production processes. Non-adherence to the terms of agreements without reward or penalty for defaulter(s) has demotivated farmers from engaging fully and committedly to sugarcane cultivation, and many smallholders have left the sector, particularly in Turiani. Generally, the lack of enforcement has led processors to ignore some of their responsibilities, even the basic ones. For example, millers delay signing the agreed contract without apparent reason. Signing of the agreement in Turiani is fixed for 30 April each year, but this date is not observed. As one respondent complained:

> Imagine this contract documentation is supposed to be in use for the operations just started,[1] yet it is not signed, and unfortunately we cannot do anything, because one time we used to go to court, but no action and or justice was observed. We finally called for the Honourable Minister for Agriculture and Food Security, but he refused to attend our concerns.

All the scenarios discussed above are summarised in Table 6.6. They bear testimony to the absence of proper contract farming arrangements between smallholders and processors and hence are hindering the potential for process upgrading.

Conclusion and policy implications

The objective of this chapter has been to assess whether contract farming was facilitating process upgrading and to identify the aspects of contract farming that enable or hinder upgrading. The study has revealed that training, input provisioning, extension services and research are not included in the contract farming arrangements, although theoretically this can be suggested because of their crucial role in the provision of smallholders' capabilities.

Table 6.6 Summary of constraints for process upgrading of smallholder sugarcane production

Key constraints	*Reasons mentioned within and across cases*
Limited incentives	Low, unfairly determined and dictated sugarcane prices
	Late in some occasions very late payments, based on unfairly quality determination payment systems
	Low production due to unreliable harvests, factory capacity limits large supply and lack of alternative markets
	Lack of enforcement and competitive regulatory framework
Limited capabilities	Limited production skills due to lack of training and extension services
	Limited access to credits (serious in Turiani)
	No subsidised input, e.g. fertiliser and late delivery of non-subsidised inputs
	Low application of modern technologies, e.g. fertiliser, due to low net returns
	Limited investment in agricultural infrastructure, e.g. irrigation facilities

Source: authors' fieldwork (2012).

Actually, the determination of purchase prices and payment modalities are the only elements included in the agreement that are theoretically predicted to be incentive factors for farmers to participate actively in the arrangement. However, the study revealed that the price of sugarcane and the payment mechanisms involved actually reduced smallholders' incentives to participate actively in sugarcane production. The procurement price was considered very low (especially in Turiani compared to Kilombero) and payments were delayed, and in some cases in Turiani extremely delayed.

Excluding training, input provision, extension services and research from the agreement weakens the capability of smallholders to adopt best practices in sugarcane farming. Given the importance of sugarcane in reducing poverty among smallholders, there is a need to develop integrated interventions that expand producer capabilities and incentives to enhance sugarcane supply. These may include the following elements:

1 Capitalising fully on the research and training institutions formerly established in Tanzania expressly to develop the sugar sector. This could be arranged by providing sufficient resources to conduct research and training aimed at strengthening the capacity of the out-growers by promoting good cane-husbandry skills and techniques related to the development of the sugarcane sector.
2 Intervening in contract farming arrangements to ensure that the state, in collaboration with other stakeholders (e.g. NGOs and CBOs), oversees the prices of both outputs and agro-inputs by setting indicative prices similar to what is the case for other crops like cotton and tobacco.
3 Revising current contract farming agreements with the purpose of making them more binding by involving stakeholders such as farmers, farmers' group leaders and associations much more in their negotiation.
4 Introducing and reviewing incentive policies and regulations to attract and retain small and larger farmers' investments in the sugarcane sector by creating alternative markets for their products. One solution could be to amend the Sugar Industry Act (2001) to allow the establishment of supplementary factories around the existing ones.
5 Retaining a government stake in the companies. Such measures could help solve the problems with the pricing and payment mechanisms, increase enforcement efficiency and reduce the excessive powers vested in the processors in the negotiations and execution of the contract. Especially in the case of MSEL, this may be a way to ease the problems that currently affect the performance and upgrading of production processes among smallholders.

Appendix 6.1 Out-grower associations included in the study

The associations located in Ruembe area and serving Kilombero 2 (K2) sugar factory include, first of all, the Ruembe Cane Outgrowers Association

(RCGA), formed on 22 May 1992, with 3,507 registered out-growers up to the date of interview (July/August 2012). This is the largest association in Ruembe area, with 72 per cent of the total volume of sugarcane delivered to the factory in aggregate of all associations in the area. The remaining 28 per cent was shared by the remaining associations, namely the Muungano Cane Growers Association (MUCGA), the Msindazi Cane Growers Association (MCGA), the Bonye Cane Growers Association (BSGA) and the Maendeleo Cane Growers Association (MACGA).

In the Kidatu area, a total of seven associations serving Kilombero 1 (K1) factory were involved in the study. First of all, the Kilombero Cane Out-growers Association (KCGA) formed on 14 June 1992, the largest association in the Kidatu area with 3,500 registered out-growers on the day of interview (July/August 2012). The remaining six associations are the Msolwa Ujamaa Cane Growers Association (MUCGA), the Mang'ula Cane Growers Association (MCGA), the Msolwa Station Nyange Cane Growers Association, the Mwangani Mtendezi Lukoga Cane Growers Association, the Muungano Cane Growers Association and the Harambee Cane Growers Association.

In Turiani, the study included the Mtibwa Outgrowers Association (MOA), which was established on 24 July 1996, the biggest cane-growers' association in Turiani division, with 6,300 registered members on the date of interview (July/August 2012). The Turiani Cane Out-growers and Other Crops Primary Cooperative Society Limited (TUCOCPRICOS LTD) is the only cooperative in the area. It was established in October 2006 with 3,000 registered members.

Note

1 The operations in 2011 started in early May.

References

Barrett, C. B., Bachke, M. E., Bellemare, M. F., Michelson, H. C., Narayanan, S. and Walker, T. F. (2012). Smallholder Participation in Contract Farming: Comparative Evidence from Five Countries. *World Development, 40*(4), 715–730.

Bolwig, S., Gibbon, P. and Jones, J. (2009). The Economics of Smallholder Organic Contract Farming in Tropical Africa. *World Development*, 37(6), 1094–1104.

Da Silva, C. (2005). *The Growth Role of Contract Farming in Agrifood Systems Development Drivers: Theory and Practice.* Working Document, Agricultural Management and Finance Services, FAO, Rome.

Emongor, A. R., Louw, A., Kirsten, F. J. and Madevu, H. (2004). *Re-governing Markets: Sourcing Small Producers' Participation in Restructured National and Regional Agri-food System.* Zambia Country Report Prepared for International Institute for Environment and Development (IIED), London, Department of Agricultural Economics, August.

Eyhorn, F., Mader, P. and Ramakrishnan, M. (2005). *The Impact of Organic Cotton on the Livelihoods of Smallholders.* FiBL Research Report. Frick: Research Institute of Organic Agriculture.

FAO (2011). FAOSTAT Agriculture Country Profiles. Online: www.fao.org/countryprofiles/index/en/?iso3=Tza Accessed 15 December 2016.

Gereffi, G. (1999). International Trade and Industrial Upgrading in the Apparel Commodity Chain. *Journal of International Economics*, 48(1), 37–70.

Grosh, B. (1994). Contract Farming in Africa: An Application of the New Institutional Economics. *Journal of African Economies*, 3(2), 231–261.

Humphrey, J. (2004). *Upgrading in Global Value Chains*. ILO Working Paper No. 28. Geneva: ILO.

Humphrey, J. (2006). *Global Value Chain in Agrifood Sector*. Working Paper, UNIDO Strategic Research and Economic Branch, United Nations Industrial Development Organisation.

Humphrey, J. and Schmitz, H. (2002). How Does Insertion in Global Value Chain Affect Upgrading in Industrial Clusters? *Regional Studies*, 36(9), 1017–1027.

Jaffee, S., Henson, S. and Rios, L. D. (2011). *Making the Grade: Smallholder Farmers, Emerging Standards and Development Assistance Programs in Africa*. Research Program Synthesis, Report no. 62324-AFR. Washington, DC: World Bank.

Kadigi, M. J., Mdoe, N. S. Y., Senkondo, E. and Mpenda, Z. (2007). *Effect of Food Safety Standards on the Livelihoods of Actors in the Nile Perch Value Chain in Tanzania*. Working Paper 2007/24. Copenhagen: Danish Institute of International Studies (DIIS).

Kamuzora, A. K. N. (2010). Contractual Governance in Agro-industry Institutions in Tanzania: A Case Study Analysis. Unpublished PhD thesis. University of Groningen.

Kaplinsky, R. and Morris, M. (2008). Value Chain Analysis: A Tool for Enhancing Export Supply Policies. *International Journal of Technology Learning*, 1(3), 283–308.

Kilombero (2014). *Outgrowers' Sugarcane Production Trend for 1999/2000–2013/2014*. Compiled report, unpublished first draft. Morogoro, Tanzania: Kilombero.

Kumar, J. (2006). Contract Farming through Agribusiness and State Corporation: A Case Study of Punjab. *Economic and Political Weekly*, 41, 5367–5375.

Matango, R. (2006). *Mtibwa Outgrowers' Scheme: A Model for Smallholder Cane Production in Tanzania*. Prepared for the UNCTAD Expert Meeting on 'Enabling Small Commodity Producers in Developing Countries to Reach Global Markets', 11–13 December. New York: United Nations Conference on Trade and Development.

Mitchell, J., Keane, J. and Coles, C. (2009). *Trading Up: How a Value Chain Approach Can Benefit the Rural Poor*. London: Overseas Development Institute.

MOA Cane Supply Agreement (2011). *Cane Supply Agreement between Mtibwa Outgrowers Association (MOA) and Mtibwa Sugar Estates Ltd., from 2009–2012*. Turiani, Morogoro.

Mtibwa (2014). *Outgrowers' Sugarcane Production Trend for 1999/2000–2013/2014*. Compiled report, unpublished first draft. Morogoro, Tanzania: Mtibwa.

Nazir, A., Jariko, G. A. and Junejo, M. A. (2013). Factors Affecting Sugarcane Production in Pakistan. *Pakistan Journal of Commerce and Social Sciences*, 7(1), 128–140.

Nkonya, N. and Barreiro-Hurle, J. (2012). *Analysis of Incentives and Disincentives for Sugar in the United Republic of Tanzania*. Technical Notes Series, MAFAP. Rome: FAO.

Prowse, M. (2012). *Contract Farming in Developing Countries: A Review*. A Savoir, Agence Française de Développement (AFD) Research Development, France.

REPOA (2012). *Transformation of the Organisation Structure for Sugarcane Outgrowers in Tanzania*. Research Report submitted to the Sugar Board of Tanzania, 1 October 2012, Research on Poverty Alleviation (REPOA), Tanzania.

SBT (2014). *Production Data: Sugar Production for 2002/03–2013/14*. Online: www.stb.go.tz/index.php/production-datas. Accessed 23 May 2014.

Setboonsarng, S., Leung, P. and Stefan, A. (2008). *Rice Contract Farming in Lao PDR: Moving from Substance to Commercial Agriculture*. ADBI Discussion Paper, No. 90. Tokyo: Asian Development Bank Institute.

Singh, S. (2005). *Contract Farming for Agricultural Development: Review of Theory and Practice with Special Reference to India*. Ahmedabad: Centre for Management in Agriculture, Indian Institute of Management.

Singh, G. and Asokan, S. R. (2005). *Contract Farming in India: Text and Cases*. New Delhi: Oxford and IBH Publishing House.

URT (2013). *Ministry of Agriculture, Food Security and Cooperatives 2012/13 Budget Speech*. Dar es Salaam: United Republic of Tanzania.

USAID (2006). *Indonesia Cocoa Bean Value Chain Case Study*. MicroREPORT, ACDI/VOCA, Microenterprise Advancement Project Business Development Services Knowledge and Practice Trade Order, United States Agency for International Development.

Young, L. M. and Hobbs, J. E. (2002). Vertical Linkages in Agri-food Supply Chains: Changing Roles of Producers, Commodity Groups, and Government Policy. *Review of Agricultural Economics*, 24(2), 428–441.

7 Coordination and upgrading in agricultural value chains

Contract farming arrangements in the Tanzanian cotton sector

Marianne Nylandsted Larsen and Paul Maganga Nsimbila

Introduction

This chapter examines the implementation of a state/donor-driven contract farming system in the cotton value chain in Tanzania. The implementation of contract farming in the late 2000s appeared after almost two decades of unrestrained competition between a large number of private buyers and ginneries, a period in which the sector struggled with fluctuating and declining crop production and yields and deteriorating quality (Tschirley *et al.*, 2010; Mwinuka and Maro, 2013; Lorenzetti, 2014). During this period, most private cotton ginneries were reluctant to procure and supply inputs to smallholders on cash or credit terms because of a combination of low levels of liquidity and high levels of competition (Gibbon, 1999; Larsen, 2006). Against this background, contract farming was promoted and coordinated by the Tanzania Cotton Board in collaboration with and financially supported by international donors. One of the main objectives was to improve coordination in the value chain and provide a sufficient supply of inputs on credit to smallholders. Besides, the establishment of contracts between smallholders organised in so-called 'farmer business groups' (hereafter FBGs) and ginneries was perceived as a means to enhance quantity and quality management and thereby upgrade the value chain by increasing farm income and export earnings from trade in cotton lint.

The chapter analyses the division of labour between two main stakeholders in the cotton chain in Tanzania – the FBGs and the cotton ginneries – in implementing contract farming by documenting some of the outcomes of contract farming during the pilot phase (2008–11). The chapter discusses to what extent improvements in chain coordination have resulted in upgrading of the chain. It is argued that value chain coordination and upgrading appear to be closely linked in this case, as improvements in coordination have facilitated product and process upgrading. However, the contractual system has been influenced by a number of key constraints. The chapter therefore points out some of the barriers to state-led upgrading interventions in (private) agricultural value chains. The chapter is organised into six main sections.

Following this introduction, the second section engages with the current conceptual debates on upgrading and value chain coordination in agricultural value chains. The establishment and organisation of contract farming in the cotton sector in Tanzania are outlined in the third section, while the fourth describes the division of labour between FBGs and ginneries in the contractual system and the outcomes of contract farming during the pilot phase. The fifth section provides some broader reflections on the fragility of the organisational set-up and incentive structures within which private contract farming in cotton has evolved in Tanzania. In conclusion, the chapter considers whether the current upgrading typologies provided by the literature on global value chains (GVC) is of any use in understanding this system.[1]

Upgrading and value chain coordination

In the GVC literature, the concept of upgrading refers to the opportunities for developing country producers to make better products more efficiently and with more skilled workers (Gibbon and Ponte, 2005). Thus, upgrading is seen as a strategy that increases returns to producer countries by injecting efficiency, productivity and skill-related inputs into the value chain. Gereffi (1999: 51–52) refers to industrial upgrading as 'a process of improving the ability of a firm or an economy to move to more profitable and/or technologically sophisticated capital and skill intensive economic niches'. In other words, 'firms can upgrade their processes ("doing things better") or products ("making better things"), or they can aim at moving to higher value-added stages in the chain like design or marketing', so-called functional upgrading (Schmitz and Knorringa, 2000: 181). Typically, the upgrading process is examined through the lenses of how knowledge and information flow within value chains from lead firms to their suppliers or buyers. Upgrading has therefore often been analysed in terms of the sequence (and implicit hierarchies) in the acquisition of functional capabilities (Gereffi, 1994). Gereffi (1999), for example, shows how certain Hong Kong clothing manufacturers learned to 'move up the value-added ladder' by designing and branding their own products while sub-contracting part of their manufacturing. This transition has occurred on the basis of organisational learning through relationships with so-called lead firms in the global clothing chain: the experience of working with the latter group of companies typically enhances technical competence and/or provides organisational knowledge to reduce production costs.

In Humphrey and Schmitz's seminal work (2002: 1020), upgrading is further defined in terms of the following categories:

- Process upgrading: firms can upgrade by transforming inputs into outputs more efficiently by reorganising the production system or introducing superior technology.
- Product upgrading: firms can upgrade by producing products that generate more value-added per employee. This involves changing the

firm's positioning in product markets. This can be achieved by changing customers or by repositioning the value chain, thus shifting the whole chain to higher-value products.

- Functional upgrading: firms can acquire new functions in the chain, such as design or marketing, or alternatively abandon existing functions to increase the overall skill content of activities. This involves changing the inter-firm division of labour within the chain or moving to another chain.
- Inter-sectoral upgrading: firms can apply competences developed in one function of a chain to a different chain or sector.

Humphrey and Schmitz (2002) emphasise that distinct types of upgrading in a particular GVC are dependent on the nature of the relationships that exist between the main actors involved in a transaction: 'different forms of chain governance have different upgrading implications' (Humphrey and Schmitz, 2002: 1023). The incentives of lead firms to facilitate the upgrading of suppliers beyond process and product upgrading (i.e. functional upgrading) require that the lead firms governing the chain 'are willing to relinquish' lower-value-added functions up- and/or downstream in the chain. In the literature on global agricultural value chains, upgrading has mainly been analysed through Humphrey and Schmitz's categories (2002), listed above, namely product, process, functional upgrading and inter-sectoral upgrading. In addition, key arguments underlying these studies are that mechanisms for upgrading should mainly be sought within the chain and that upgrading possibilities in a particular GVC are influenced by the nature of the relationships (governance) that exist between the main actors in the value chain (see, for example, Dolan and Humphrey, 2004; Lee *et al.*, 2010).

The notion of upgrading proposed by Humphrey and Schmitz (2002) has been criticised on several points. For example, Ponte and Ewert (2009) question the ability of their rather rigid taxonomy of types of upgrading to capture upgrading processes in the 'real world'. The empirical case Ponte and Ewart discuss in making this point is upgrading in the South African wine sector, which they analysed using Humphrey and Schmitz's categories. The authors point out the empirical difficulties of distinguishing between product and process upgrading, especially in agro-food products (Ponte and Ewert, 2009: 1638). For example, the introduction of new processes (e.g. husbandry practices at field level) generates new categories of products (e.g. organic products) with increased unit value (product upgrading). In addition, Ponte and Ewert provide a detailed discussion of upgrading possibilities in agro-food GVCs, including processes usually treated as 'downgrading' in the GVC literature, such as selling lower-value products in higher volumes. They interrogate the implicit perception that upgrading implies a fixed series of stages (i.e. from process to chain upgrading) and suggest that upgrading is understood as 'reaching a better deal including a balance between rewards and risk' (Ponte and Ewert, 2009: 1637). This broader understanding of upgrading as 'reaching a better deal' and the importance of studying concrete patterns that seem to offer suppliers higher and more stable returns were initially suggested by Gibbon (2003) based on empirical observations of the global coffee and cotton chains. Gibbon (2003) suggests analysing the 'reward structures' available to suppliers in global value chains, as well as 'upgrading preconditions and mechanisms'. Reward structures refer to rewards or opportunities available to suppliers, for example, how price premiums and discounts are determined in the international market for a particular crop (see also Gibbon, 2001). Bringing 'upgrading preconditions and mechanisms' into the discussion stresses the importance of empirical observations of the particular national and local circumstances that influence suppliers' ability to generate higher and more stable rewards.

Research has also criticised the lack of broader development issues in how upgrading is conceptualised (Bair, 2008, 2005; Tokatli, 2012). For example, Bair (2005) emphasises the problem of 'scaling out' from processes of upgrading at the firm level to regional or national development. This problem remains, despite recent efforts to develop closer links between upgrading and the territorial dimension (Fold, 2014). In addition, Bair (2005) and Tokatli (2012) ask 'upgrading for whom?' Both authors point out that the upgrading of suppliers (e.g. suppliers acquiring new functions) in GVCs does not necessarily result in, for example, improved working conditions. For producers in developing countries, mechanisms and opportunities for upgrading are mainly sought from suppliers within the chain and mediated through ties with global lead firms. However, Tokatli (2012) stresses the importance of analysing successful upgrading by suppliers that is not initiated or driven by lead firms or mediated through firm–firm relationships. The capabilities and strategies of suppliers in upgrading and/or their relationships with non-economic actors like the state are pivotal to analyses that aim to advance the understanding of real-world patterns of upgrading. As Tokatli also emphasises, a focus on suppliers themselves 'would considerably enrich the literature' on upgrading and contribute to a broader understanding of upgrading, including upgrading patterns influenced by chain 'external' dynamics (Tokatli, 2012: 11).

While most of the GVC literature focuses on upgrading in GVCs initiated and driven by private capital and on the role of buyers (lead firms) and chain governance in defining upgrading opportunities, this chapter analyses a recent state/donor-led initiative aimed at restoring the link between smallholders, processers and remunerative export markets by improving coordination along the value chain and providing opportunities to upgrade the value chain. It analyses the role of the state in facilitating upgrading and the responsibilities of two groups of private actors in the cotton value chain – the FBGs and private ginneries – in implementing the state-initiated contract farming system. Finally, the chapter discusses whether the current typologies of upgrading discussed above are useful in understanding the state-led upgrading of agricultural value chains.

Development of a contractual system in the Tanzanian cotton sector

Until the onset of economic reforms in the early 1990s, the cotton sector in Tanzania was characterised by widespread state regulation and intervention, although the instruments and agents used changed from time to time. Cotton marketing was organised through a three-tier system consisting of primary cooperative societies, regional cooperative unions and the parastatal Cotton Board (later renamed the Tanzania Cotton Board, hereafter TCB). The latter performed overall regulatory functions and was responsible for sales of cotton lint on both the domestic and export markets. Regional cooperative unions (RCUs) had a legal monopoly in purchasing and ginning all seed cotton, while primary cooperative societies (PCSs) purchased seed cotton on behalf of the RCUs. Chemicals (mainly insecticides) were imported by the TCB, advanced to the RCUs on credit against future lint sales and then distributed to the farmers through the PCSs, likewise on credit (Dercon, 1993; Undolle, 1997). The liberalisation of the cotton sector in the early 1990s and the combination of state withdrawal and low entry barriers into the cotton trade and ginning resulted in a highly competitive market in which the private sector rapidly came to dominate the country's cotton production. Most of the new private buyers constructed their own ginneries, the number of which (and hence ginning capacity) increased significantly. In the 2011/12 marketing season, 43 ginneries were operating as opposed to around 24 in the early 1990s (TCB, 2013). Competition between ginners to obtain market share included competing on geographical coverage and the availability of cash during the season, as well as on price. In order to secure sufficient supplies of seed cotton to run their ginneries at full or at least sufficient capacity, it was quite common for most ginneries to rely on commissioning private buying agents, who operated over extremely far-flung areas (Larsen, 2008).

In this context, the relationship between the two main groups of actors in the value chain, smallholders and ginneries, was 'arm's-length' in kind, basically involving only the exchange of seed cotton based on market principles. With the introduction of competition, farmers regularly began to obtain higher shares of the world market price than previously, and were paid promptly in cash for their seed cotton. However, competition in the seed cotton market undermined or placed considerable constraints on the links between input supply on credit and output marketing by greatly increasing the scope for side-selling by farmers. Just a few seasons after liberalisation, the RCUs stopped providing inputs on credit, while the availability of inputs on cash terms dropped dramatically, as only a few private input traders entered the market (Gibbon, 1999; Larsen, 2008). This almost certainly contributed to lower yields and to the general decline in average national production in the late 1990s and early 2000s (Mwinuka and Maro, 2013; Larsen, 2006). Average seed cotton yields (kg/ha) have ranged between 560 and 750 kg of seed cotton per hectare (around 220 and 300 kg per acre), with significant downward trends at the end of the 1990s and in the mid-2000s. Fluctuations in yields are driven by weather conditions (drought, floods, heavy rainfall associated with hailstones), poor germination of cotton seeds and insufficient quantities of different types of chemicals during the season, their use being inappropriately timed (interview with TCB, Mwanza and Bunda, January 2013).

As illustrated in Figure 7.1, seed cotton production is characterised by huge inter-seasonal variations, which, especially in years of low or lower production, has created a 'scramble for cotton' among ginners in order to minimise their unit ginning costs and fulfil forward supply contracts. As a result, the quality control and grading system collapsed: ginneries and buying agents followed a strategy of buying seed cotton regardless of quality, for which there was no feedback mechanism and therefore no direct incentives for farmers to deliver better-quality cotton. The quality of cotton has remained a critical factor in the global cotton value chain, and as a consequence of deteriorating quality, Tanzanian lint lost all premiums on the world market (Lorenzetti, 2014; Tschirley *et al.*, 2010; Larsen, 2008).[2]

Several new interventions were undertaken by the TCB to provide a minimal level of crop protection chemicals to farmers, such as the 'passbook' and 'voucher' systems of the 2000s. These initiatives were based on levies paid by the private sector and coordinated by a crop fund (the Cotton Development Fund, later renamed the Cotton Development Trust Fund, hereafter CDTF) involving representatives of both public institutions and the private sector. The passbook and voucher systems were both beset by severe

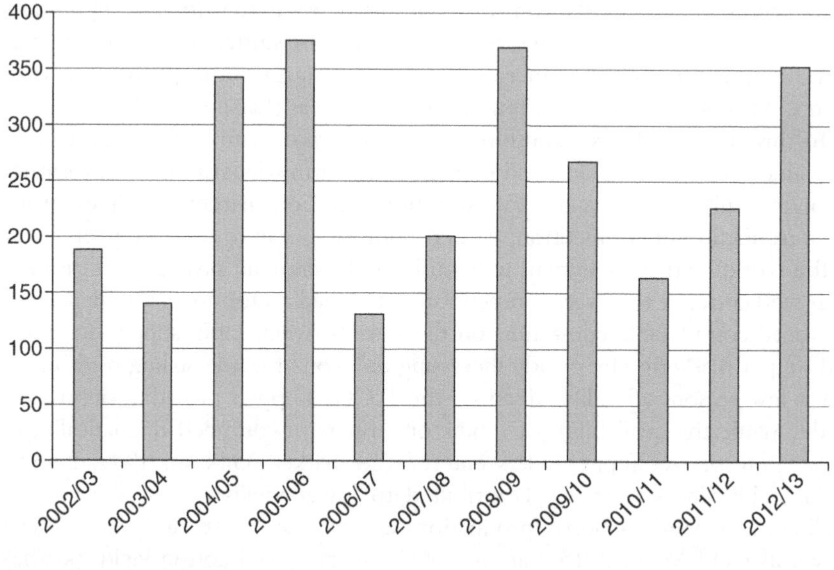

Figure 7.1 Seed cotton production in Tanzania, 2002–13.

Source: TCB (2013).

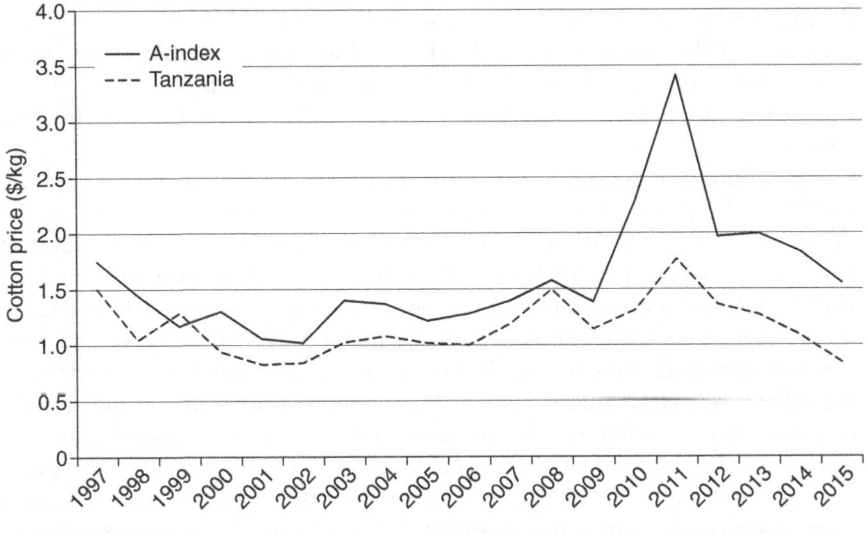

Figure 7.2 Lint unit prices, A index and Tanzania, 1997–2015.

Source: IMF (2016).

problems and subsequently abandoned. Against this background, in 2009 a 'Cotton Sector Development Strategy' was jointly formulated by the key stakeholders in the sector, to be supported financially by international donors, notably the Tanzania Gatsby Fund.[3] In addition, a cotton and textile development programme has recently been drafted to focus not only on improving cotton yields and the quality of cotton lint, but also on developing the textile industry in Tanzania (Tanzania Gatsby Trust, 2011; Mwinuka and Maro, 2013; TCB, 2010). As part of the Cotton Sector Development Strategy, contract farming was promoted and coordinated by the TCB in collaboration with the Tanzania Gatsby Fund and TechnoServe.

TechnoServe, which is supported financially by the Tanzania Gatsby Fund, drafted the contents of the contract on behalf of the TCB. The draft contract was subsequently reviewed and commented on by the two main stakeholders: the ginneries through the Tanzania Cotton Association and smallholder farmers through the Tanzania Cotton Growers Association (TACOGA), and both parties made proposals for improvements to the contract. In principle, all sections of the contract were agreed by stakeholders during the 2008/09 Annual General Meeting of the Tanzania Cotton Association, while legal officers from the cotton-growing districts were involved in reviewing and amending the contract. The contract does not have a clause on seed cotton prices. Instead, an indicative seed cotton price (for 'contracted' and 'non-contracted' seed cotton) is settled annually through a stakeholders' consultative meeting and announced by the TCB before each marketing season. The so-called floor price is based on prevailing world market prices for lint according to the A Index (see note 2), estimates of the prices of oil and cake, and agreed applicable levies, taxes and ginning and marketing costs in the case of the ginners, while production costs, applicable levies and taxes are taken into account for the farmers.

The contract is primarily a buying-cum-input credit agreement, and the CDTF is the principal agency responsible for the overall assessment of the inputs required in the WCGA and the bulk procurement and distribution of inputs to ginneries. The TCB's District Office, in collaboration with the District Task Force, estimates the inputs required at the village level according to the acreage expected to be cultivated by members of FBGs as well as non-contract cotton farmers in the district. Input requirements are submitted to the CDTF, which in turn is responsible for the importation and distribution of inputs to the ginneries. The contract stipulates that the ginnery agrees to purchase the so-called minimum input packages from the CDTF. The 'input package' provided to smallholders is defined by the CDTF as one minimum input package per acre – the so-called acre packages. An acre package contains cotton seeds and three bottles of chemicals to be applied per acre. The recommended spraying regime is six sprays per season per acre for optimal effect, with three sprays per season seen as the absolute minimum, hence the term 'minimum' input packages.

Price competition between private buyers and ginners remained a political imperative. It was therefore agreed to base the contractual scheme on a 'competitive model' to ensure that smallholders continued to benefit from price competition during the marketing season. The cotton-growing districts were divided into geographically restricted zones, with only three ginneries being licensed to operate in each zone in order to increase price competition, but also to restrain competition as a means of avoiding side-selling by smallholders. The TCB is responsible for 'overseeing' the number of buyers present in each zone during the marketing season to protect the interests of both parties.

The contract farming scheme was piloted in Bunda district (Mara region) between 2008 and 2011. Five ginneries were licensed and involved in contract farming, and around 26,000 farmers of an estimated 35,000 to 40,000 smallholder cotton-growers organised in groups were provided with inputs on credit by the ginneries (interview with TCB, Mwanza and Bunda, January 2013), which were allowed to purchase seed cotton from both contract and non-contract farmers (interviews with TCB and TechnoServe, Bunda, January, 2013). In addition to training FBGs and ginneries in input delivery and quality management procedures, TechnoServe and a District Task Force facilitated annual 'touring around campaigns' to link FBGs to specific ginneries. Competing ginneries 'stood shoulder to shoulder' in the cotton villages and campaigned to attract FBGs through the provision of other services than input credit as part of the ginneries' competitive assets.

Tightening value chain coordination through contract farming

The role of FBGs in contract farming

The revival of the PCSs as organising devices for smallholder cotton-growers was apparently not politically important. Instead, the TCB and TechnoServe acknowledged the need to develop new-generation (i.e. post-liberalisation) farmers' organisations. As mentioned above, to gain access to inputs on credit, smallholders were required to join FBGs, established the main entity between the seed cotton farmers and the ginneries in the contractual system, with a legal mandate to enter into business contracts with the ginners. The contract signed by FBGs and ginneries specifies the roles and responsibilities regarding input provision on credit. In the design phase of the FBGs, smallholders were required to form groups of 60–100 members. Once registered as a group, the members were trained by TechnoServe staff in democratic procedures to elect a board (i.e. a chairman, a secretary and council member) to be responsible for the FBG's activities. In addition, TechnoServe provided training to the FBG board members in keeping accounts and their obligations related to the contractual arrangements with the ginneries.

Three main tasks were allocated to the FBGs. First, they are responsible for distributing inputs (seeds and crop-protection chemicals) on credit to smallholder cotton farmers. Prior to the cultivation season, the FBG board is required to collect expected acreages from FBG members in order to estimate the required inputs. When inputs are delivered to the FBGs by the ginneries at the beginning of the cultivation season, the FBGs are responsible for distributing inputs to member farmers and keeping records of the distributions to recover input loans from farmers at the time of the cotton sales.

Second, it was expected that FBGs would provide a more efficient platform for the provision of extension services. In the pilot phase of contract farming, the concept of the 'Training of Trainers' was launched and financially supported by the Cotton Sector Development Programme. So-called lead farmers (strongly performing contract cotton farmers) selected from within established FBGs were trained in good agricultural practices such as conservation agriculture and insect pest management. Further, training in how to guide and disseminate improved farming practices and management to other FBG group members was provided (interviews TCB, Tanzania Gatsby Fund and TechnoServe, Mwanza and Bunda, January 2013).

Third, the FBGs were expected to be responsible for purchasing seed cotton from member farmers and conducting quality control on behalf of the ginneries. The critical motivation for the TCB and TechnoServe to locate the responsibility for primary purchases and quality control at the FBG rather than the ginnery level was to shorten the chain between smallholders and ginners and 'squeeze' private buying agents out of the value chain. TechnoServe staff actively assisted the FBGs in quality-control procedures at the point of primary purchase and provided extensive training in keeping accounts. The FBGs could decide whether they wanted to become engaged in primary purchases of seed cotton on behalf of the ginneries and as such 'take on board' a portfolio of new buying agent tasks such as quality and input recovery control, the monitoring of member farmers and logistical issues. If the FBG decided not to function as a buying agent during the marketing season, its main tasks then became weighing and keeping records of distributed inputs on credit and delivering cotton seed, as well as transferring the bulk cotton to buying agents and/or to the ginnery's buying outlet in the area.

The incentive for becoming engaged in primary purchases is based on an agreed commission to be paid by the ginneries, partly to cover overheads and partly to enhance the financial assets of the FBG for the benefit of its members. The FBG board was given a mandate to set up a secretariat responsible for purchasing seed cotton and appointing several individuals from among themselves to assume different roles, namely a clerk, an assistant clerk and a security guard. In addition to the FBG's ordinary tasks of weighing and record keeping, the FBG assumed responsibility for financial matters such as collecting working capital from the ginnery. In addition, the FBG was expected to arrange the transportation of seed cotton to the ginnery. During quality control and weighing at the ginnery, and in cases where the seed cotton is found to be wet or to contain contaminated material, the clerk is responsible for offloading the consignment in special areas for drying and/or removing the contaminated material or unsuitable seed cotton and then reloading it for sale. An agreed commission of 30 Tsh per kg of purchased seed cotton was provided, of which the clerk, assistant clerk and security guard were given 15 Tsh per kg seed cotton purchased from farmers for operational management, including a subsistence allowance for the clerk to escort the loaded vehicle to the ginnery for a final quality test. The remaining balance (15 Tsh per kg seed cotton) was intended to increase the FBG's financial assets.

The role of ginneries in contract farming

The ginneries were envisaged as providing inputs on credit to FBGs and as purchasing a majority of their seed cotton intake directly from smallholders through contracted FBGs concentrated in fairly small geographical zones near the ginneries. Hence, the costs of monitoring and sorting seed cotton at the factory would be considerably reduced, as the initial quality control was to be situated at the FBG level. Furthermore, the overheads resulting from operating own-employment buying posts and the risks associated with both employee and buying agent fraud were minimised. Staff members of TechnoServe provided technical assistance and training in disseminating procedures for efficient input delivery systems and book-keeping for input credit recovery purposes (interviews with ginneries and TechnoServe, Bunda, January 2013).

The ginneries were 'licensed' to distribute inputs on credit to FBGs in the restricted zones when the company had signed contracts with 35 to 50 FBGs (depending on the size of the groups) and had deposited 50 million Tsh with CDTF for input purchases (interviews with CDTF and TCB, Mwanza and Bunda, January 2013). In order to increase the availability of crop-protection chemicals and cotton seeds for those smallholders who were not engaged in contract farming in the restricted zones, the ginneries were asked to supply input packages on cash terms as well.

As mentioned above, the CDTF and local public agencies are responsible for determining the required volume and types of chemicals to be imported and the logistical aspects of the distribution system. Crop-protection chemicals and chemicals for seed treatment are distributed to the ginneries according to the yearly estimates of input requirements, while the ginneries are responsible for distributing inputs to both FBGs (on credit term) and individual smallholders (on cash terms) in a timely fashion according to the cotton cultivation calendar. The ginneries keep records of inputs provided on credit to the FBGs to ensure deduction of costs during the marketing season. The ginneries are only obliged to provide a minimum input package per acre; based on crude estimates of the costs of seed cotton production where the recommended quantities of seeds (10 kg) and pesticides (six bottles) per acre are issued, the minimum input package provided by ginneries on credit covers between 3 and 7 per cent of total costs (for elaboration, see Nsimbila, 2015).

Outcomes of contract farming in the pilot phase

As explained above, the TCB and TechnoServe have played a pivotal role in improving coordination of the cotton value chain through the organisation of cotton smallholders in FBGs and the establishment of contracts between FBGs and ginneries. Fieldwork revealed a number of positive and negative outcomes of contract farming in the pilot phase, which will be considered in turn in this section.

On the one hand, four positive outcomes of the contractual system and direct linkages between FBGs are worth noting. First, during the pilot phase, the ginneries have gradually increased the number of FBGs under contract farming. One of the largest ginneries in terms of market share started to provide inputs on credit to 47 FBGs in Bunda District (Ginnery2, Table 7.1), a figure that had increased to 316 FBGs by 2012/13 by covering FBGs in other districts in the WCGA. This ginnery purchased around 70 per cent of seed cotton from FBGs and 30 per cent from private buying agents. Hence, it was still common for the ginneries to use more than one purchasing system, for example, purchasing seed cotton through commissioned buying agents, as well as from FBGs. In a few cases, 'buying agents come in at a late stage in the marketing season and are used as a "last resort" in order to fulfil ginning capacity' (interview with ginnery, Bunda, January 2013). In addition, some of

Table 7.1 Number of FBGs per ginnery in Bunda district, 2009–12

	FBGs 2009/10	*FBGs 2010/11*	*FBGs 2011/12*
Ginnery1	72	74	74
Ginnery2	47	47	84
Ginnery3	52	43	52
Ginnery4	–	–	29
Ginnery5	–	–	91
Total	171	164	300

Source: authors' fieldwork (2013).

the ginneries in Bunda District started to offer land preparation services (renting out tractors) to FBG members partly on credit terms: contract farmers had to pay 30,000 Tsh per acre up front and in cash for the service, while a further 10,000 Tsh were deducted from crop sales during the marketing season. In a few cases extension services were systematically tied to provision of input credit to increase the productivity of inputs (interviews with ginneries, TCB and TechnoServe, Bunda, January, 2013). In most cases where FBGs had taken on the role of buying agents (9 of the 16 FBGs interviewed), they have remained with the same ginnery since the establishment of contract farming in the pilot phase. These FBGs have also obtained extension and tractor services from the ginnery more frequently, most notably in the first couple of years in the pilot phase.

Second, in most cases the FBGs revealed significant increases in production and yield levels as a result of direct linkages between FBGs and ginneries and improved access to inputs on credit and other services provided by the contracted ginnery. For example, the secretary of one of the FBGs interviewed emphasised that before the implementation of contract farming, he harvested between 200 and 400 kg per acre, but after the FBG had formed and inputs on credit supplied to its members, the yield rose to 800 kg per acre (interview with FBG, January 2013). More broadly, the results of the farmer survey suggest that contract farmers achieved on average 260 kg per acre, while non-contract farmers obtained 232 kg per acre. Nonetheless, mean cotton yields in Tanzania are still low relative to many rain-fed cotton-producing countries in sub-Saharan Africa (see e.g. Tschirley *et al.*, 2010).

Third, input repayment rates and quality control were also higher when FBGs were operating as buying agents than in the cases of private buying agents purchasing seed cotton from FBGs or non-contract farmers (interviews with FBGs and ginneries, Bunda, January 2013). The main reason for this was that the clerk, assistant clerk and secretariat were part of the FBG, and they monitored and supervised its members in repaying their input loans (i.e. not side-selling). Also, these FBGs more frequently rejected seed cotton contaminated with sand, water, stones and other non-cotton materials in order to decrease the costs associated with the removal of contaminated material found during quality checks at the ginnery. Hence, by placing the responsibility for initial quality control with the FBG, the quality concerns and associated price discounts in international markets have to some extent been addressed.

Fourth, the availability and application of crop-protection chemicals distributed by ginneries against cash payments or on credit terms increased steadily during the pilot phase of the contractual system. As illustrated in Table 7.2, both contract farmers and non-contract farmers in the farmer survey applied more than five 'one-acre packages' in the 2011/12 production season. This indeed suggests that the use of agro-chemicals and the number of sprayings applied by both contract and non-contract farmers have increased significantly relative to earlier years in the post-liberalisation period (see, for example, Gibbon, 1999; Larsen, 2006; Poulton *et al.*, 2004).

On the other hand, the contractual system has been influenced by a number of key constraints, of which four crucial aspects are worth noting. First, smallholders and ginneries are both vulnerable to the vagaries of the international lint market, and the resilience of the contract farming system is challenged by price movements on international markets. For example, seed cotton prices became a major bone of contention in the 2011/12 marketing season. The marketing season started at a time when world market prices were at their highest in recent years at around US$2 per pound of lint (see Figure 7.2), and the floor price was settled at 1,100 Tsh per kg of seed cotton. Within a very short time, however, lint prices in international markets declined to less than US$1.20 per pound, resulting in new negotiations over the floor price. After prolonged negotiations, a consensus was finally reached with a floor price of 800 Tsh per kg of seed cotton. None of the ginneries managed to purchase seed cotton at that price, and farmers saw prices fall from 1,100 to 660 Tsh per kg of seed cotton at the beginning of the marketing season. Smallholders started 'free-riding' on the ginneries' input provision, and the ginneries obtained very low credit recovery rates.

Second, and in relation to the last point, a more fundamental issue concerns the lack of mechanisms to minimise side-selling to other ginneries and/or private buying agents. During the first couple of years of contract farming, the ginneries in Bunda District obtained fairly high input credit-recovery rates (between 60 and 80 per cent). In the 2010/11 and 2011/12

Table 7.2 Comparison of mean measures for contract and non-contract farmers

	Contract farmers (N=211)	*Non-contract farmers (N=189)*
Average acres	4	3.11
Average pesticides used per acre (in acre packages)	5.5	5.30
Average cotton seeds used per acre (kg)	12.12	10.60
Average yield per acre (kg/acre)	260	232

Source: authors' fieldwork (2012).

marketing seasons, FBGs' repayment rates declined markedly, partly because of a steep fall in prices on the international lint market, as indicated above, but also due to intensified competition in the restricted zones. In 2010/11, complaints concerning a lack of sufficient price competition to improve seed cotton prices at the village level became more strident, and a rural member of parliament pressured the Ministry of Agriculture, Food Security and Cooperative to allow one more buyer (ginnery) to enter the zones in Bunda District to improve price competition. In addition, one 'roaming buyer' not licensed by the TCB or the ministry was able to purchase seed cotton in the zones for a period of a month without any repercussions from the board. The entry of additional buyers in the zones caused a slight increase in the seed cotton price. At the same time, these buyers had sufficient liquidity (working capital) to make instant payments to farmers, and contract farmers started defaulting on their repayments of inputs to the ginneries, though several FBGs repaid some of the loans by selling seed cotton to other buyers and paying in cash to the contracted ginnery. Yet, the provision of other services by ginneries such as extension services and land preparation declined or stopped completely, presumably because the ginneries found that competitors reaped the benefits when smallholders started side-selling their cotton production.

Third, the input delivery system is organised 'externally' to the cotton value chain, that is, through the CDTF and local public agencies, and the availability of different types of chemicals during the season still seemed to be a problem. The organisational aspects of the input delivery system influence the ginneries' ability to fulfil their obligations to distribute chemicals and seeds in a timely fashion according to the cotton cultivation calendar. For example, in the 2011/12 season there was a delay in estimating the quantity of cotton seeds and pesticides needed in Bunda District. As a result, the CDTF ordered dressing chemicals for seed treatment too late, which resulted in late preparation and treatment of cotton seeds at the ginneries and land preparation on the part of the farmers. Late estimates on the amounts of pesticides required also delayed the ordering, importation and distribution of pesticides to ginneries and hence to smallholders. These failings obviously influence the efficiency of chemical applications to counter outbreaks of disease and pest infestations during the season. In addition, interviews with key informants, ginneries and smallholders revealed that many farmers are still unfamiliar with water-based chemicals (imported by CDTF) as opposed to oil-based pesticides, which were the main type of chemical available several years ago, which, according to some farmers, 'stuck to plants whenever sprayed and killed pests effectively' (interviews with contract farmers, Bunda, January 2013).

Fourth, while FBGs were perceived to play a pivotal role in delivering well-sorted, uncontaminated seed cotton to the ginneries, incentives to deliver better-quality cotton are not reflected in a price premium attached to sorted, high-quality seed cotton at the primary marketing stage. Financial incentives for raising the quality of seed cotton only apply to farmers who have opted for contract farming, and they require the FBG board to be willing to purchase seed cotton on behalf of the ginnery. Many FBGs decided not to function as buying agents due to the associated operational risks and uncertainties but preferred instead simply to record and bulk the seed cotton received from FBG members. These FBGs had fewer incentives to enforce quality controls when collecting seed cotton from smallholders. In addition, private buying agents are still common, and the quality of seed cotton from them seemed to be lower (i.e. more contaminated) than supplies from FBGs. In cases where ginners rely on buying agents to purchase seed cotton from centres in the villages, both smallholders and buying agents frequently contaminated the seed cotton by adding water, sand, oil, stones and other non-cotton materials to it in order to increase its weight. Hence, the contractual system and the ginneries' purchasing practices (using both buying agents and FBGs) do not provide clear and consistent incentives to smallholders (whether contract or non-contract farmers) to deliver better-quality cotton.

Besides, none of the FBGs that also operated as buying agents had received an annual commission from the ginneries based on the volume of traded seed cotton. This was constantly raised by FBGs during interviews in 2013, and the issue of a 'lacking commission' was a battlefield between the FBGs and ginneries. On the other hand, the ginneries emphasised that, because of increased side-selling by smallholders, the agreed commission had to cover some of the costs of FBG members defaulting on their loans. In addition, one of the premises of the contract arrangements was that FBGs only purchase seed cotton from member farmers, though in fact they were increasingly buying seed cotton from both contract and non-contract farmers. As the commission is based on the total volume of seed cotton traded by the FBGs, the ginneries raised the financial burden involved as a serious drawback of the contract farming arrangements.

Concluding remarks

This chapter has explored upgrading in the value chain for cotton since the implementation of contract farming in the late 2000s. The Tanzania Gatsby Trust Fund and the international NGO TechnoServe facilitated the implementation of a contract farming system in the cotton sector by providing both finance and technical support to the TCB. Contract farming was implemented in Bunda District during a pilot phase, and five ginneries entered into contracts with newly established farmer FBGs. The contract farming scheme was based on voluntarism: smallholders could choose to become part of the scheme and obtain inputs on credit from the ginneries, though they had to be a member of an FBG. Further, the FBGs could engage in the trading of seed cotton. Contract farming was also based on a competitive model in order to ensure that smallholders continued to benefit from price competition during the marketing season. The analysis in this chapter indicates that enhanced coordination in the value chain through contract farming has improved efficiency in the input credit delivery system and quality management, thus contributing positively to both product and process upgrading. However, the fragility of the organisational arrangements and incentive structures within which private contract farming has evolved in Tanzania is threatening the viability of the contractual arrangements between FBGs and ginneries.

Returning to the discussion of concepts of upgrading in agricultural GVCs, this study reveals promising signs of product and process upgrading as a result of improvements in the coordination of the cotton value chain. Hence, by directing attention to vertical coordination and facilitating the establishment of direct linkages between FBGs and ginneries through contract farming, the efficiency of the input credit delivery system has been significantly enhanced. This improvement seems to result in both process and product upgrading: with regard to the former, the management of the timely delivery of inputs on credit to smallholders can be characterised as process upgrading, narrowly defined as transforming inputs into outputs more efficiently. However, more efficient management of the input delivery system also contributes to improvements in the quality of seed cotton and cotton lint (product upgrading), as the timely application of pesticides reduces cases of 'damaged', downgraded lint caused by infestations of pests and insects. For example, one of the main problems concerns the so-called cotton stainer, a fungus disease that stains the lint yellow and weakens the fibre. This is a common cause of the downgrading of lint in international markets, in addition to the lower yields obtained at the farm level through reduced seed weights.

The reorganisation of quality management (process upgrading) and the implementation of initial quality control at the FBG level point towards product upgrading and the likelihood of improvements in unit prices on the international markets, though it was not possible to discover the magnitude of product upgrading measured in the classic form of higher unit prices in international markets. Taken together, improvements in value chain coordination and upgrading patterns appear to be intimately related. However, it is difficult to discern a particular hierarchy in upgrading processes (e.g. does product upgrading follow from process upgrading?), while enhancing the coordination of inputs on credit to FBGs appears to result in both product and process upgrading taking place simultaneously.

Finally, the extent to which state-led interventions to improve coordination in the cotton value chain have resulted in chain upgrading remains an open question. Widespread side-selling and breaches of contracts by both smallholders and ginneries surfaced in Bunda District before the rolling-out of the contract farming scheme to the entire WCGA. Stakeholders felt that a lack of ownership of the entire 'contract farming project', combined with a credit default crisis, subsequently led to the collapse of the TCB/donor-driven contract farming system after it was 'rolled out' in 2012 covering more than 50 competing ginneries and around 10,000 newly established FBGs. Side-selling has ended all attempts to provide inputs and other services on credit to smallholders on a larger scale since the collapse of the contract farming arrangement, though a more 'silent', privately driven contract farming system has emerged in Bunda District in the last couple of years. This might have clear consequences for social differentiation within cotton communities, and perhaps also for regional differentiation, though the medium- and longer-term effects of these dynamics remain to be seen. This in turn points to the crucial role of the specific national and local circumstances in shaping the prospects for value chain upgrading, as Gibbon suggested (2003). A different set of chain-upgrading strategies would need to take into account the nature of private competition and low production capacity utilisation in the ginnery segment of the chain in Tanzania.

Notes

1 The chapter is based on findings from a survey of 400 smallholders (211 contract farmers and 189 non-contract farmers) conducted in Bunda District in the Western Cotton Growing Area (WCGA) in 2012. This was complemented with qualitative in-depth interviews conducted with 16 FBGs and five ginneries licensed to operate contract farming in Bunda District in 2013. In addition, in-depth interviews were conducted with contract farmers in the same FBGs. Interviews were also carried out with representatives of the Tanzania Cotton Board, the Cotton Development Trust Fund, TechnoServe, the Tanzania Gatsby Fond, the Tanzania Cotton Growers Association and government staff (extension officers) at the district and ward levels. For a description of fieldwork methods, see Nsimbila (2015).

2 There are a variety of pricing systems for world cotton, but in general the A Index is considered to be the most authoritative. The A Index is an index of the level of offer prices on the international market based on an average of the cheapest five quotations from a selection of 16 countries, including Tanzania (Larsen, 2008). Tanzanian lint has been discounted by US$0.015 to 0.02 per pound relative to the A Index since the mid-1990s due to its low quality.

3 The Tanzania Gatsby Trust is mainly funded by the World Bank, the Italian government, the Kilimo Trust, the Ford Foundation and DANIDA (Tanzania Gatsby Trust, 2011).

References

Bair, J. (2005). Global Capitalism and Commodity Chains: Looking Back, Going Forward. *Competition & Change*, 9(2), 153–180.

Bair, J. (2008). Analysing Global Economic Organization: Embedded Networks and Global Chains Compared. *Economy and Society*, 37(3), 339–364.

Dercon, S. (1993). Peasant Supply Responses and Macroeconomic Polities: Cotton in Tanzania. *Journal of African Economies*, 2(2), 157–194.

Dolan, C. and Humphrey, J. (2004). Changing Governance Patterns in the Trade in Fresh Vegetables between Africa and the United Kingdom. *Environment and Planning A*, 36, 491–509.

Fold, N. (2014). Value Chain Dynamics, Settlement Trajectories and Regional Development. *Regional Studies*, 48(5), 778–790.

Gereffi, G. (1994). The Organization of Buyer-Driven Global Commodity Chains: How US Retailers Shape Overseas Production Networks. In G. Gereffi and M. Korzeniewicz (eds), *Commodity Chains and Global Capitalism* (pp. 93–123). Westport, CT: Greenwood Press.

Gereffi, G. (1999). International Trade and Industrial Up-grading in the Apparel Commodity Chain. *Journal of International Economics*, 48, 37–70.

Gibbon, P. (1999). Free Competition without Sustainable Development? Tanzania Cotton Sector Liberalisation, 1994/95 to 1997/98. *Journal of Development Studies*, 36(1), 128–150.

Gibbon, P. (2001). Upgrading Primary Production: A Global Commodity Chain Approach. *World Development*, 29(2), 345–363.

Gibbon, P. (2003). *Commodities, Donors, Value Chain Analysis and Upgrading*. Paper presented at the United Nations Conference on Trade and Development (UNCTAD).

Gibbon, P. and Ponte, S. (2005). *Trading Down: Africa, Value Chains & the Global Economy*. Philadelphia, PA: Temple University Press.

Humphrey, J. and Schmitz, H. (2002). How Does Insertion in Global Value Chains Affect Upgrading in Industrial Clusters? *Regional Studies*, 36, 1017–1027.

IMF (International Monetary Fund) (2016). IMF Primary Commodity Prices. Online: www.imf.org/external/np/res/commod/index.aspx Accessed 21 September 2016.

Larsen, M. N. (2006). Market Coordination and Social Differentiation. *Journal of Agrarian Change*, 6(1), 102–131.

Larsen, M. N. (2008). The International Cotton Market and Cotton Sector Reforms in Sub-Saharan Africa. In N. Fold and M. N. Larsen (eds), *Globalization and Restructuring of African Commodity Flows* (pp. 156–183). Uppsala: The Nordic Africa Institute.

Lee, J., Gereffi, G. and Beauvais, J. (2010). Global Value Chains and Agrifood Standards: Challenges and Possibilities for Smallholders in Developing Countries. *PNAS Early Edition, 2010*, 12326–12331. Online: www.pnas.org/cgi/doi/10.1073/pnas.0913714108 Accessed 18 December 2011.

Lorenzetti, A. L. (2014). Unravelling the Magnitude of Sub-Saharan Africa Cotton Quality in Sector Reform Outcomes. *Revista Internazionale di Scienze Sociali*, 2014(4), 401–430.

Mwinuka, L. and Maro, F. (2013). *Analysis of Incentives and Disincentives for Cotton in the United Republic of Tanzania*. Rome: MaFAP, FAO, Technical Notes Series (1–48).

Nsimbila, P. M. (2015). The Role of Contract Farming on Productivity of Smallholder Cotton Producers in Bunda District and Effects on Cotton Value Chain in Tanzania. Unpublished PhD Thesis. Mzumbe University, Tanzania.

Ponte, S. and Ewert, J. (2009). Which Way Is 'Up' in Upgrading? Trajectories of Change in the Value Chain for South African Wine. *World Development*, 37, 1637–1650.

Poulton, C., Gibbon, P., Hanyani-Mlambo, B., Kydd, J., Larsen, M. N., Maro, W., Osorio, A., Tschirley, D. and Zulu, B. (2004). Competition and Coordination in Liberalized African Cotton Market Systems. *World Development*, 32(3), 519–536.

Schmitz, H. and Knorringa, P. (2000). Learning from Global Buyers. *Journal of Development Studies*, 37, 177–204.

Tanzania Gatsby Trust (2011). *The Annual Report 2011*. Online: http://tanzania-gatsby.com/v3/annual-reports. Accessed 5 December 2016.

TCB (2010). *Second Corporate Strategic Plan: 2010/11–2012/13*. Online: www.cotton.or.tz/second-corporate-strategic-plan-201011-201213. Accessed 5 December 2016.

TCB (2013). *Annual Report and Accounts for the Year Ending 30th June 2013*. Online: www.cotton.or.tz/tanzania-cotton-board-annual-report-2013. Accessed 5 December 2016.

Tokatli, N. (2012). Toward a Better Understanding of the Apparel Industry: A Critique of the Upgrading Literature. *Journal of Economic Geography*, 2012, 1–19. Doi:10.1093/jeg/lbs043.

Tschirley, D. L., Poulton, C., Gergely, N., Labaste, P., Baffes, J., Boughton, D. and Estur, G. (2010). Institutional Diversity and Performance in Africa Cotton Sectors. *Development Policy Review*, 28(3), 295–323.

Undolle, P. M. (1997). *Marketing Review of Cotton, 1995/96*. Dar es Salaam: Ministry of Agriculture, Marketing Development Board.

Part III

Contract farming and household economics

8 Tobacco contract farming in the Urambo District of Tanzania

Which farmers obtain inputs on credit and which buy them for cash?

Bahati M. Ilembo, Joseph Andrew Kuzilwa and Arne Henningsen

Introduction

Tobacco is a remunerative cash crop for smallholder farmers in Urambo District, Tanzania, as well as in many other regions of the world. However, tobacco production requires intensive use of purchased inputs, particularly agrochemicals. Most smallholder tobacco farmers do not have sufficient liquid financial resources for purchasing these inputs and, thus, can only acquire these inputs if they can obtain them on credit. Based on a cross-sectional data set of 300 smallholder tobacco contract farmers, this chapter analyses which tobacco farmers obtain the inputs on credit and which buy them for cash.

This chapter starts with a description of the broader background of our analysis, followed by a section about the methodology and a section that presents the results. The final section presents our conclusions.

Background

Smallholder farmers' access to credit

Increased production and productivity of the agricultural sector is seen as an essential element of any strategy to stimulate the economic growth of any productive region (Konlambigue, 2007; UNDP/URT, 2015). However, an increase in agricultural production will largely depend on the adoption of more intensive technologies. For any kind of business, adequate financial capital, through own savings, loans or both, is necessary to buy what is needed to produce goods or services. The adoption of more intensive technologies entails increased financial capital for buying not only intermediate inputs, but also better equipment. In agricultural production, expenditure on inputs usually occurs during the planting and growing period of the crops, while returns are received several months later after harvest. In this respect, to finance the purchase of inputs, farmers must either use their savings or obtain credit (Diagne *et al.*, 2000). As most smallholder farmers in developing countries have little or no liquid financial resources, access to credit can significantly increase their ability to acquire agricultural inputs (Barrett *et al.*, 2012). Hence, credit is one of the most important drivers in farmers' production and access to it can substantially accelerate the adoption of modern agricultural technologies and production patterns because it is an important gateway for financing inputs, such as improved seeds and inorganic fertilisers, which can enhance agricultural productivity (Swaminathan *et al.*, 2010). Credit can also help farmers to smooth their consumption when they face income shocks (Zhu and Li, 2007). Agricultural credit is normally provided in cash, but some programmes provide in-kind loans for agricultural inputs such as seed, fertiliser and other farm production inputs. The provision of in-kind loans is a strategy to reduce the fungibility of the loan. Smallholder farmers' demand for agricultural credit is likely influenced by their own financial resources, the returns on their potential investments, their attitude towards the risk of borrowing, transaction costs on the credit market (e.g. the complexity of administrative procedures) and the conditions of the offered loans (e.g. interest rate, repayment schedule).

Numerous studies show that many smallholder farmers all over the world have constrained access to credit, particularly in less developed countries (see, e.g., Okten and Osili, 2004; Brau and Woller, 2004; Hu *et al.*, 2011). This restricted access to credit ranges from partial exclusion in developed countries to full exclusion in less developed countries (Hu *et al.*, 2004). Because of market failures and the perceived riskiness of agricultural production, formal financial institutions have been reluctant to provide loans to smallholder farmers and, if they do it, they have stringent lending terms and conditions, which exclude most smallholder farmers from borrowing. As an alternative, farmers sometimes borrow money to finance their agricultural production through informal lending channels, but these informal lending arrangements often claim higher interest rates and, thus, often leave farmers in debt and trapped in poverty (Hu *et al.*, 2011). This has been confirmed by studies that show that credit markets in developing countries work inefficiently due to a number of reasons: first, the existence of monopoly powers in credit markets normally exercised by informal lenders; second, high transaction costs incurred by borrowers during loan application and processing; and third, the moral hazard problem (e.g. Carter, 1988; Bell *et al.*, 1997). If these factors combine, most smallholder farmers find themselves out of the loan market. Although credit is often found to be a determining factor in profits and investments, multiple market failures on the credit market may overwhelm the effect of credit (Diagne and Zeller, 2001). Farmers' limited access to credit hinders the intensification of agriculture production.

Contract farming

Contract farming arrangements have the potential to facilitate the modernisation of poor smallholder farmers' production technology because they address several challenges faced by smallholder farmers. For instance, by reducing the transaction costs associated with bargaining and enforcement, contract farming can address failures on markets for credit, inputs, and outputs and, thus, provide easy access to credit, agricultural inputs, such as seeds and fertilisers, and a guaranteed market for the output (Grosh, 1994; Key and Runsten, 1999). It may also solve critical problems of access to information, technical assistance and extension services (e.g. Kirsten and Sartorius, 2002; Melese, 2010). Barrett *et al.* (2012) compare contract farming participation in five countries in Africa and note that limited access to credit is an institutional constraint that reduces the attractiveness of participating in contract farming for smallholders. This finding emphasises the importance of the provision of credit in contract farming arrangements.

Tobacco contract farming and input credit in Tanzania

Since Tanzania liberalised agricultural input supply in 1994, the trade with and provision of agricultural inputs, to a large extent, has been left to private companies. This was when tobacco-buying companies started to enter into contracts with cooperative societies. Cooperatives are legally established associations of persons who have voluntarily joined together for the purpose of achieving a common goal and who accept the risks and the benefits of this undertaking. The cooperatives act as intermediaries between agribusiness firms and smallholder farmers, which forms a contracting model popularly known as 'firm + intermediary + smallholder'. According to the Tanzanian Tobacco Act 2001, only farmers with a tobacco production contract are allowed to sell tobacco. Smallholder farmers can voluntarily become a member of a cooperative society in order to join the tobacco contract farming arrangement. All farmers who want to become a member of the cooperative society need to sign a one-page contract and commit themselves to selling all of their produced tobacco to the contracted buyer at the quantity and quality described in the purchase contract between the buyer and the primary cooperative society. The registration for the tobacco contract farming arrangement is for a particular production season, but it can be renewed annually. Virtually all smallholder tobacco farmers in Tanzania have a tobacco production contract with a so-called primary cooperative society, of which they are a member. Larger tobacco producers can be 'independent' and have a direct production contract with a tobacco trader, but smallholder farmers usually cannot meet the requirements for 'independent' tobacco farming.

The tobacco traders have specific requirements regarding tobacco production (e.g. application of specific inputs) and the curing process, which should make sure that the produced tobacco leaves meet the required quality standards. In order to facilitate this, tobacco contract farming arrangements not only usually provide a reliable outlet for the produced tobacco, but also extension services and the required inputs on credit. Initially, tobacco-buying companies supplied inputs on credit to the society's members who committed themselves to selling their produced tobacco to the cooperative society and repaying the loan after having sold their tobacco. However, since 2009, the inputs are supplied directly by the cooperatives (e.g. Guo and Jolly, 2008), i.e. the primary cooperative societies provide an 'input package' to their members on credit and subtract the costs of the 'input package' from the revenues from the tobacco that the farmers deliver to the cooperative society. The 'input package' consists of the approved seeds, recommended agrochemicals (i.e. inorganic fertilisers and pesticides) and other inputs such as packaging materials (e.g. jute twines, hessian cloth). Farmers who choose not to receive this 'input package' (on credit) from their cooperative society need to buy the inputs for tobacco production from other sellers, such as commercial input traders, who usually sell the inputs for cash, but sometimes supply the inputs to the farmers on credit. Hence, the tobacco farmers have up to three options for obtaining the inputs for tobacco production: (1) obtain them from their cooperative society on credit; (2) obtain them from other sellers on credit; or (3) buy them from other sellers for cash.

The first option, i.e. obtaining the 'input package' on credit from the cooperative society, demands that the farmer is a member of a cooperative society (which all smallholder tobacco farmers need to be) and joins a group of five to ten other members of the same primary cooperative society. All farmers in a group are jointly responsible for the repayment of the loans that were given to its members, i.e. if one or more members of a group do not repay (a part of) their loan, the other group members need to step in and repay the non-repaid loans of the defaulting group members ('group guarantee'). The farmers who would like to obtain inputs on credit from the cooperative society need to sign an additional one-page contract that is popularly called an 'input credit contract'.

The second option, i.e. obtaining inputs on credit from other sellers, is difficult to choose for many smallholder farmers because private input traders do not usually sell inputs on credit, or if they do, they have strict requirements for this.

The third option, i.e. buying inputs from other sellers for cash, demands that farmers have sufficient liquid financial resources or access to cash loans from, for example, private money-lenders or friends. Only a few farmers have sufficient liquid financial resources for buying all the inputs for tobacco production for cash. Furthermore, cash loans from private money-lenders usually have high interest rates and many farmers do not have friends or relatives who have sufficient liquid financial resources and are willing to give them loans.

In the following, we discuss a few potential advantages and disadvantages of the three above-mentioned options. According to information that we obtained during qualitative interviews, the prices that the cooperative societies charge for the inputs are very similar to those other sellers charge for them so that input prices are not an important criterion for farmers when they choose between the three options.

During the qualitative interviews, we also heard from farmers that at least some of them have access to various sources for loans to finance their agricultural activities. Apart from formal financial institutions, farmers could obtain loans from private money-lenders or friends based on established trust and networks. Money-lenders who live in the same village as their borrowers can visit their clients and gain first-hand information about their credit-worthiness. This helps to reduce transaction costs and build trust and it increases the chance that farmers can obtain loans. In these cases, reputation within networks may serve as a kind of social collateral for loans (Wydick *et al.*, 2009). Farmers can also use their social networks to obtain information on various options for financing agricultural production without depending on input credit obtained from a contract farming system. In our qualitative interviews, we also heard that the interest rates charged by commercial input sellers, private money-lenders and friends are very diverse and may be lower or higher than the interest rates charged by the cooperative societies.

During our qualitative interviews, we heard that many tobacco farmers do not like the group guarantee for the repayment of the loans. Tobacco farmers who want to form or join a group usually thoroughly assess the characters of other potential group members in order to minimise the risk of having to repay the loans of defaulting group members. In spite of this, individual farmers have frequently defaulted on the repayment of their loans and it has been a significant problem in every season (although the proportion of defaulting farmers differed between cooperative societies). The non-repayment of loans usually occurs when a tobacco farmer delivers insufficient tobacco (or no tobacco at all) to his/her cooperative society so that its monetary value is too little to repay the input loan. This can be caused by, for example, adverse weather conditions, inappropriate agricultural practices or so-called side-selling to other tobacco buyers. From our qualitative interviews, we learned that most tobacco farmers do not like the group guarantee because they do not like to be held responsible for other group members' potential failures or misconduct. Therefore, some smallholder tobacco farmers chose not to obtain the input package on credit from the cooperative society, but instead attempted to obtain inputs on credit from other sellers who do not require a group guarantee or tried to gather cash from personal savings or (informal) loans in order to buy inputs for cash from other sellers.

Finally, during our qualitative interviews, some farmers claimed that they prefer to obtain their inputs from commercial sellers because sometimes the cooperative societies do not deliver the inputs on time, which results in their delayed application, thereby potentially reducing their effectiveness and, thus, productivity and profitability.

Adoption of input credit

A number of theoretical and empirical models have been proposed to explain farmers' use of credit for buying agricultural inputs. Meyers (1982) argues that while adoption is an individual decision, it is nonetheless influenced by the social context. In particular, Meyers (1982) introduced what he called the 'structure of opportunity', which makes the farmers' decision not entirely autonomous. The theoretical framework is grounded in the sociological premise that social stratification, i.e. differential possession and/or access to resources, is a critical determinant of a farmer's ability and willingness to adopt credit. This fits well with the case of tobacco production in Tanzania, which is mostly conducted by smallholder farmers with different resource limitations and different income levels. Combining technological and sociological factors, Meyers (1982) analysed the socio-economic determinants of credit adoption for agricultural households in a semi-arid district of Kenya. His analysis included households that operated in the same agro-ecological zone and used similar agricultural technologies, but some of these households had one or more loans, while the other households did not have any loans. Meyers (1982) found that credit adoption varied substantially with income levels and credit history. In particular, he found that households with a medium income that depended on crop income and households with a more educated household head were more likely to use credit than households that depended on non-farm income, households that were relatively rich and households that employed multiple strategies to earn income. The study also found that households with low income adopted credit for non-economic reasons.

Recently, Sebatta *et al.* (2014) used a double-hurdle model of credit access (for the opportunity structure and actual use) and studied factors that determine the decisions of smallholder agricultural farmers in Zambia to access rural finance and the intensity of their participation in the financial markets. Using a sample of 1,300 farm households, the study found that the level of education of the head of household positively influenced credit adoption, while the size of the household negatively influenced credit adoption.

Methodology

Data collection

The empirical study is based on a cross-sectional data set obtained from a survey of tobacco contract farmers conducted in the Urambo District of Tanzania between August and September 2012. A structured survey questionnaire was prepared and administered for the collection of the data.

Multistage sampling was used where, in the first stage, we systematically selected ten primary cooperative societies from a list of 85 that was obtained from the district council. The selection of the primary cooperative societies was based on the geographical set-up (whether in urban or rural areas) and the buyer they served (whether Alliance One or Tanzania Leaf Tobacco) in order to have some diversity of the selected primary cooperative societies, both regarding the geographical set-up and regarding the tobacco buyer. The sampling frame for the survey consisted of a list of all members of the selected primary cooperative societies, i.e. only tobacco farmers that were members of a cooperative society were included, while independent farmers were not included because there were too few to include them in a quantitative analysis. In total, 300 tobacco contract farmers were selected using simple random sampling from the list of all members of the selected primary cooperative societies by using a table of random numbers. In instances where a randomly selected respondent was not available, the next available tobacco farmer was selected. The survey collected information on several variables including socio-economic variables, crop-specific information (production and sales), information about the institutional environment, as well as individual and demographic factors. In order to supplement the quantitative data and facilitate the interpretation of the results, we also obtained qualitative information through open questions.

Conceptual framework

In general, an individual who aims to take up a loan first needs to know where he or she can borrow the money before he or she can decide to apply for a loan (Okten and Osili, 2004). This is somewhat different for input credit for tobacco contract farmers in Tanzania because they need to be a member of a primary cooperative society that offers an input package on credit and encourages their farmers to obtain it. Therefore, it is reasonable to assume that all farmers in our sample have access to inputs on credit and are aware of this opportunity. In this case, our empirical analysis does not need to take into account the sequential nature of the adoption of input credit by explicitly modelling the availability or the awareness of inputs on credit (e.g. with a double-hurdle model) so that it can focus on the question of whether farmers obtain inputs on credit or buy them for cash.

As we want to analyse farmers' adoption of input credit, we distinguish between two groups of farmers: those who obtain at least some of their inputs on credit (either from the cooperative society or from other sellers) and those who bought all their inputs for cash (only from other sellers because the cooperative societies do not sell inputs for cash). We conceptualise the farmers' choice between these two mutually exclusive options by way of a microeconomic model. Let $U_i(Y_i, X_i)$ indicate the utility function of farmer i, where Y_i is a binary variable that denotes whether farmer i obtained inputs on credit ($Y_i = 1$) or bought inputs for cash ($Y_i = 0$) and X_i is a vector of other explanatory variables. The farmer chooses to obtain inputs on credit if this implies a higher utility level than buying the inputs for cash, i.e. $U_i(Y_i = 1, X_i) > U_i(Y_i = 0, X_i) \Leftrightarrow U_i(Y_i = 1, X_i) - U_i(Y_i = 0, X_i) > 0$. Consequently, the probability that a farmer i chooses to adopt an input credit can be written as: $\text{Prob}(Y_i = 1) = \text{Prob}(U_i(Y_i = 1, X_i) > U_i(Y_i = 0, X_i))$.

Econometric model specification[1]

As our econometric analysis investigates farmers' choice between two distinct mutually exclusive options, the dependent variable of our analysis (Y_i) is a binary variable. Estimating the effects of explanatory variables on a binary dependent variable by standard linear econometric models, such as Ordinary Least Squares (OLS), results in the so-called linear probability model. Even though the linear probability model may be appropriate under certain circumstances, it is problematic in most cases because the disturbance terms of these models are likely heteroscedastic, while the predicted probabilities are not necessarily bounded between 0 and 1 (Pindyck and Rubinfeld, 1981; Greene, 1994). Hence, when analysing the determinants of a binary dependent variable, it is usually advantageous to use an econometric method that takes the binary nature of the dependent variable into account, for example, the Logit or Probit models (e.g. Gujarati, 1988). In our analysis, we apply the Logit model in order to investigate the probabilities of the farmers' choice between 'obtain inputs on credit' and 'buy inputs for cash' depending on a set of the explanatory variables (X_i).

The Logit model is specified as $Y_i^* = \beta X_i + \mu_i$, where $Y_i^* \equiv U_i(Y_i = 1, X_i) - U_i(Y_i = 0, X_i)$ is an unobserved latent response variable, β is a vector of parameters to be estimated, X_i is a vector of explanatory variables and μ_i is the error term. While Y_i^* is unobservable, we observe the dummy variable Y_i, which is defined as:

$$Y_i = \begin{cases} 1 & \text{if } Y_i^* > 0 \text{ (i.e. farmer } i \text{ adopted the input credit)} \\ 0 & \text{otherwise} \end{cases}$$

Accordingly, the probability of input credit adoption is denoted as:

$$\text{Prob}\left(Y_i = 1\right) = \text{Prob}\left(Y_i^* > 0\right) = \text{Prob}\left(\mu_i > -\beta X_i\right) = 1 - F\left(-\beta X_i\right) = F\left(\beta X_i\right),$$

where $F(.)$ is the cumulative distribution function (CDF) of the error term μ_i. The Logit model assumes that the error term μ_i follows a logistic distribution. Hence, according to the Logit model, the probability that a farmer obtains inputs on credit can be expressed as:

$$\text{Prob}\left(Y_i = 1 | X_i\right) = \frac{1}{1 + e^{-\beta X_i}} = \frac{e^{\beta X_i}}{1 + e^{\beta X_i}}$$

In this model, the parameters β indicate the marginal effects of the explanatory variables on the latent dependent variable Y_i^* but not the marginal effects on the conditional probability $\text{Prob}(Y_i = 1)$, which are given as $\partial \text{Prob}(Y_i = 1) / X_i = \beta\ \text{Prob}(Y_i = 1)\ \text{Prob}(Y_i = 0)$ and indicate the rate of change in the probability as a result of a unit change in the explanatory variable (Mukherjee *et al.*, 1998).

Explanatory variables

In this section, we discuss the explanatory variables that we use to analyse smallholder farmers' adoption of input credit. In order to obtain consistent estimates, we made sure that all explanatory variables are exogenously given or at least pre-determined, i.e. our explanatory variables do not include any endogenous decision variables, for example regarding the production in the season that is covered by our survey. In the following, we describe all explanatory variables that we use in our econometric analysis and we explain the rationales for including them in our analysis:

a Household's workforce (*HHWF)*: The number of household members of working age, i.e. persons aged between 15 and 64 years (URT, 2013). The larger the number of household members of working age, the higher the household's income-generating capacity. On the one hand, this could increase the household's ability to buy the inputs for cash. On the other hand, this could increase the household's confidence in being able to repay the loan. Hence, it can either increase or decrease the probability of adopting input credit.
b Age of the household head (*AGEHH*, AGE^2): The age of the household head may be related to financial resources and risk attitudes. As this relationship may be non-linear, we include the squared age of the household head as an additional explanatory variable.
c Education of the household head (*EDU2*, *EDU3*, *EDU4*): These binary variables indicate the household head's level of education: variable EDU2 is 1 if the household head has a primary education and it is 0 otherwise; variable EDU3 is 1 if the household head has a secondary education and it is 0 otherwise; variable EDU4 is 1 if the household head has a tertiary education and it is 0 otherwise; if the household head has no formal education, variables EDU2, EDU3 and EDU4 are all zero. The household head's level of education indicates human capital and, thus, the ability to acquire and process information. Education facilitates the household head's capability to acquire, process and apply information, for example, related to the input credit so that households with a more educated household head could be more likely to understand and adopt input credit. Hence, this variable is assumed to have a positive effect on the adoption of input credit.
d Tobacco farming experience (*FEXP*): This variable indicates the number of years the household head has cultivated tobacco. Experience in tobacco farming may be related to the household head's risk perception and his/her confidence in obtaining a sufficiently high profit from tobacco production that allows the household to repay the loan. Experience in tobacco farming can either increase or decrease the probability of adopting input credit.
e Total farm size in acres (*TFASIZE*): This indicates the total farm land owned by the household in acres. Farm size is an indicator of wealth and a proxy for social status and influence within a community. It can either increase or decrease the probability of adopting input credit.
f Type of walls (*WMCBL*): This variable indicates the type of walls of the house in which the household lives (1 = concrete blocks, 0 = mud). This is an indicator of the household's wealth. On the one hand, wealthier

households may be less likely to adopt input credit as they may have sufficient financial resources to buy the inputs for cash. On the other hand, poor households may be reluctant to adopt input credit because they have a higher risk of not being able to repay the loan (e.g. in case of very low yields due to unfavourable weather conditions).

g Other cash income (*CASHINC*): This is a binary variable that indicates whether the household derives cash income from non-tobacco activities, i.e. wage employment, off-farm businesses as well as livestock production and non-tobacco crop production, excluding income received in kind. On the one hand, farm households with higher cash income from other activities are more likely to have greater confidence in being able to repay the loan. On the other hand, farm households with higher cash income from other activities may have sufficient financial resources to buy the inputs for cash. Hence, it can either increase or decrease the probability of adopting input credit.

Results

Descriptive statistics

Table 8.1 presents descriptive statistics of the explanatory variables that are described in the previous section. This is done separately for the adopters of input credit, for the non-adopters and for all farmers in the sample. Furthermore, Table 8.1 presents the results of statistical tests for independence of the explanatory variables and the adoption of input credit (χ^2-tests for equal proportions and t-tests for equal means). These tests indicate that adopters and non-adopters of input credit do not significantly differ regarding the workforce of their household or the age, education or tobacco-farming experience of the household head. However, adopters of input credit are significantly wealthier regarding land ownership and type of walls than non-adopters, while the proportion of households with cash income from non-tobacco activities is significantly higher among non-adopters than among adopters of input credit.

Estimation results

Table 8.2 presents the maximum likelihood estimates of the Logit model described above. A likelihood ratio test clearly rejects the null hypothesis that the parameters of all explanatory variables (except for the constant) are jointly equal to zero. This indicates that the explanatory variables are jointly significantly associated with the probability of the adoption of input credit.

In contrast to the descriptive statistics and the tests for independence in Table 8.1 that indicate correlations between the adoption of input credit and the explanatory variables, the estimation results of the Logit model in Table 8.2 indicate *ceteris paribus* effects of the explanatory variables on the adoption of input credit. However, the results of the Logit model point in the

Table 8.1 Descriptive statistics for adopters and non-adopters

	Adopters ($N_1 = 2,460$)		*Non-adopters ($N_2 = 54$)*		*All respondents ($N = 300$)*		*Statistical tests for independence*
	Count	*Proportion*	*Count*	*Proportion*	*Count*	*Proportion*	
Education (EDU[X])							
No formal schooling	26	10.6%	3	5.6%	29	9.7%	$\chi^2 = 2.79$
Primary education	202	82.1%	47	87.0%	249	83.0%	P-val. = 0.425
Secondary education	17	6.9%	3	5.6%	20	6.7%	
Tertiary education	1	0.4%	1	1.90%	2	0.7%	
Type of walls (*WMCBL*)							
Mud	213	86.6%	39	72.2%	252	84.0%	$\chi^2 = 11.66$
Concrete blocks	33	13.4%	15	27.8%	48	16.0%	P-val. = 0.003
Other cash income (*CASHINC*)							
No	179	72.80%	51	94.4%	230	76.7%	$\chi^2 = 11.63$
Yes	67	27.2%	3	5.6%	70	23.3%	P-val. = 0.001
	Mean	*Std. Dev.*	*Mean*	*Std. Dev.*	*Mean*	*Std. Dev.*	
Workforce (HHWF)	7.6	3.7	8.0	3.4	7.7	3.7	$t = 0.602$ P-val. = 0.548
Age (AGEHH) [years]	42.1	12.5	41.0	12.5	41.9	12.5	$t = -0.563$ P-val. = 0.574
Tobacco farming experience (FEXP) [years]	15.2	11.6	13.8	10.7	15.0	11.5	$t = -0.799$ P-val. = 0.425
Farm size (TFASIZE) [acres]	6.0	3.7	7.1	5.6	6.2	4.1	$t = 1.784$ P-val. = 0.076

Source: authors' fieldwork (2012).

Table 8.2 Maximum likelihood estimates of the Logit model

Variable	*Coefficient*	*Std. error*	*P-value*	*Marginal effect*
Constant	2.246	1.946	0.248	
HHSIZE	–0.012	0.051	0.817	–0.002
AGEHH	0.018	0.087	0.839	0.003
AGE^2	–0.0001	0.0009	0.906	–0.00002
EDU2	–0.920	0.658	0.163	–0.137
EDU3	–0.857	0.929	0.356	–0.128
EDU4	–2.072	1.610	0.198	–0.310
FEXP	0.016	0.021	0.447	0.002
TFASIZE	–0.095	0.043	0.027	–0.014
WMCBL	–1.016	0.389	0.009	–0.152
CASHINC	2.081	0.662	0.002	0.311

Source: authors' fieldwork (2012).

Notes

Log likelihood value = –124.296; number of observations = 300; LR test: Chi-squared value = 34.25, 10 degrees of freedom, P-value = 0.0002; Pseudo R-squared value = 0.1211. The marginal effects have been calculated at the mean values of the explanatory variables.

same direction as the tests for independence: the household's workforce and the age, education and tobacco-farming experience of the household head are not statistically significantly associated with the adoption of input credit, while the total farm size, the type of the wall and cash income from non-tobacco production are significantly associated with the adoption of input credit. Owning one additional acre of land decreases the probability of adopting input credit by around 1.4 percentage points; the probability that a household which lives in a house with walls built of concrete blocks adopts input credit is *ceteris paribus* around 15 percentage points lower than the corresponding probability of a household which lives in a house with walls made of mud; and the probability that a household with non-tobacco cash income adopts input credit is *ceteris paribus* around 31 percentage points higher than the corresponding probability of households without non-tobacco cash income.

Conclusion

This chapter has investigated the factors that are associated with input credit adoption among smallholder tobacco farmers in the Urambo district of Tanzania. The analysis is based on a cross-sectional data set that was collected through a farm-household survey from 300 smallholder tobacco farmers who are registered members of a cooperative society. The results of our Logit model indicate that human capital variables (i.e. the household's workforce and the age, education and tobacco-farming experience of the household head) are not significantly associated with the adoption of input credit, while wealth indicators (i.e. land ownership and type of walls) are *ceteris paribus* significantly negatively associated with the adoption of input credit and the availability of non-tobacco cash income is *ceteris paribus* significantly positively associated with the adoption of input credit. Thus, our results suggest that wealthier households which own relatively large areas of farm land and live in a house with walls made of concrete blocks tend to have access to sufficient financial resources to purchase inputs for cash, while households which have cash income from non-tobacco activities *ceteris paribus* (e.g. having the same wealth indicators such as total land ownership and type of wall) are more confident that they can repay the loan than households who do not have cash income apart from tobacco production.

We suggest that future research takes a closer look at some aspects that we have discovered in this study. For instance, one could distinguish between more than two groups of farmers, for example, farmers who obtain inputs on credit, farmers who obtain cash loans for buying inputs and farmers who use their own financial resources for buying inputs for cash. One could also distinguish between the different reasons farmers have for not adopting input credit, such as having sufficient liquid financial resources themselves, obtaining lower interest rates when taking cash loans or being afraid of not being able to repay the loans.

Note

1 The econometric equations and representations used in this study are derived from Hill *et al.* (2008) and Mariano *et al.* (2012).

References

Barrett, B. C., Bachke, M. E., Bellemare, M. F., Michelson, H. C., Narayanan, S. and Walker, T. F. (2012). Smallholder Participation in Contract Farming: Comparative Evidence from Five Countries. *World Development*, 40(4), 715–730.

Bell, C., Srinivasin, T. N. and Udry, C. (1997). Rationing, Spillover and Interlinking in Credit Markets: The Case of Rural Punjab. *Oxford Economic Papers*, 49, 557–587.

Brau, J. C. and Woller, G. M. (2004). Microfinance: A Comprehensive Review of the Existing Literature. *Journal of Entrepreneurial Finance*, 9, 1–26.

Carter, M. R. (1988). Equilibrium Credit Rationing of Small Farm Agriculture. *Journal of Development Studies*, 28, 83–103.

Diagne, A. and Zeller, M. (2001). Access to Credit and Its Impact on Welfare in Malawi. IFPRI Research Report No. 116.

Diagne, A., Zeller, M. and Sharma, M. (2000). *Empirical Measurements of Households' Access to Credit and Credit Constraints in Developing Countries: Methodological Issues and Evidence*. Washington, DC: International Food Policy Research Institute.

Greene, W. H. (1994). *Econometric Analysis*, 2nd edn. Englewood Cliffs, NJ: Prentice-Hall.

Grosh, B. (1994). Contract Farming in Africa: An Application of the New Institutional Economics. *Journal of African Economies*, 3(2), 231–261.

Gujarati, D. N. (1988). *Basic Econometrics*, 2nd edn. New York: McGraw-Hill.

Guo, H. and Jolly, R. W. (2008). Contractual Arrangements and Enforcement in Transition Agriculture: Theory and Evidence from China. *Food Policy*, 33, 570–575.

Hill, R. C., Griffiths, W. E. and Lim, G. C. (2008). *Principles of Econometrics*, 3rd edn. New York: Wiley.

Hu, B., Li, X. and Gan, C. (2011). The Welfare Impact of Microcredit on Rural Households in China. *The Journal of Socio-Economics*, 40, 404–411.

Hu, D., Reardon, T., Rozelle, S., Timmer, P. and Wang, H. (2004). The Emergence of Supermarkets with Chinese Characteristics: Challenges and Opportunities for China's Agricultural Development. *Development Policy Review*, 22(5), 557–586.

Key, N. and Runsten, D. (1999). Contract Farming and Smallholders and Rural Development in Latin America: The Organization of Agroprocessing Firms and the Scale of Outgrower Production. *World Development*, 27(2), 381–401.

Kirsten, J. and Sartorius, K. (2002). Linking Agribusiness and Small-Scale Farmers in Developing Countries: Is There a New Role for Contract Farming? *Development Southern Africa*, 19(4), 503–529.

Konlambigue, A. (2007). Financing Agricultural Inputs in Northern Togo through an Inventory Credit System: An Economic Analysis of Institutional Response. AAAE Conference Proceedings, 119–124. Online: http://purl.umn.edu/52100. Accessed 30 January 2017.

Mariano, M. J., Villano, R. and Fleming, E. (2012). Factors Influencing Farmers' Adoption of Modern Rice Technologies and Good Management Practices in the Philippines. *Agricultural Systems*, 110, 41–53.

Melese, A. T. (2010). *Contract Farming in Ethiopia: An Overview with Focus on Sesame.* Ethiopia: Development Cooperation, Ministry of Home Affairs.

Meyers, L. R. (1982). *A Sociological Approach to Farming Systems in Kenya.* Cornell International Agriculture Mimeograph No. 64, Department of Rural Sociology, Cornell University. Online: https://library.wur.nl/isric/fulltext/isricu_i6371_001.pdf. Accessed 28 October 2016.

Mukherjee, C., White, H. and Wuyts, M. (1998). *Econometrics and Data Analysis for Developing Countries.* London and New York: Routledge.

Okten, C. and Osili, U. O. (2004). Social Networks and Credit Access in Indonesia. *World Development*, 32(7), 1225–1246.

Pindyck, R. and Rubinfeld, D. (1981). *Econometric Models and Economic Forecasts*, 2nd edn. New York: McGraw-Hill.

Sebatta, C., Wamulume, M. and Mwansakilwa, C. (2014). Determinants of Smallholder Farmers' Access to Agricultural Finance in Zambia. *CCSE Journal of Agricultural Science*, 6(11), 63–73.

Swaminathan, H., Du Bois, R. S. and Findeis, J. L. (2010). Impact of Access to Credit on Labour Allocation Patterns in Malawi. *World Development*, 38(4), 555–566.

UNDP/URT (2015). *Tanzania Human Development Report 2014: Transformation for Human Development.* Economic and Social Research Foundation.

URT (2013). *Population Distribution by Age and Sex.* United Republic of Tanzania. Online: http://ihi.eprints.org/2169/1/Age_Sex_Distribution.pdf. Accessed 1 February 2017.

Wydick, B., Hayes, H. K. and Kempf, S. H. (2009). Social Networks, Neighbourhood Effects, and Credit Access: Evidence from Rural Guatemala. *World Development*, 39(6), 975–976.

Zhu, X. and Li, Z. (2007). Heterogeneous Impact of Farmer Credit: An Empirical Investigation Based on IVQR Model. *Systems Engineering – Theory & Practice*, 27(2), 68–75.

9 Income diversification of small-scale sugarcane contract farmers in Kilombero and Turiani, Tanzania

Thobias E. Nsindagi, Jennifer K. Sesabo and Arne Henningsen

Introduction

This chapter examines the factors that influence diversification of incomes among smallholder sugarcane contract farmers in Kilombero and Turiani in Tanzania. The main questions of interest are (i) what sorts of income-generating activities do sugarcane-growers pursue, and (ii) what influences the levels of income obtained from the various income-generating activities? To answer these questions, we base our analysis on the framework developed by Schwarze and Zeller (2005).

In Tanzania, sugarcane is a crop with a potential for socio-economic development because it creates an outlet for the production of over 30,000 farming households and offers employment to over 14,000 people in the sugarcane-processing industry (Nkonya and Barreiro-Hurle, 2012). Likewise, the sector provides about US$28 million in earnings annually and contributes about 1.7 per cent to total tax revenues (SBT, 2007; Nkonya and Barreiro-Hurle, 2012). Thus, the sector contributes to the realisation of Tanzania's Developing Vision 2025, which, among other things, focuses on improving food security and reducing rural poverty.

Given the sector's role in the economy, contract farming (CF) was introduced to promote sugarcane production. Several studies indicate that CF can enhance productivity and, hence, food security, facilitate the participation of small-scale farmers in local and global value chains, and improve the incomes and reduce the poverty of participating farmers (e.g. World Bank, 2008, 2011). Conversely, some authors argue that farmers participating in CF arrangements in developing countries face many challenges, such as low prices, delayed payments, heavy taxes, food price increases, inefficiencies in the processing industries and exploitation by large agribusiness firms due to unequal bargaining power (e.g. Little and Watts, 1994; Nkonya and Barreiro-Hurle, 2012). Such challenges may lead to stagnation in output as well as declining productivity and reducing farmers' incomes (Nkonya and Barreiro-Hurle, 2012).

In Tanzania, CF arrangements for sugarcane have been practised for many years. Experience shows that there have been fluctuations in sugarcane production in Kilombero and that sugarcane production in Turiani has declined. In Kilombero, both yields and the amount of land under sugarcane cultivation have fluctuated significantly. The declining trend in Turiani was due to a decrease in the land being used for sugarcane cultivation, as many sugarcane farmers ceased sugarcane production on some or all of their land and used it to produce other food crops instead (SBT, 2014). We assume that smallholder sugarcane-growers respond to such trends by adjusting their livelihood strategies in accordance with the sustainable livelihoods approach. Given the declining profitability of sugarcane production, many farm households aimed to improve their livelihoods by reducing sugarcane production and adopting additional income-generating activities or increasing other currently practised productive activities (Reardon *et al.*, 2006).

Any initiative aiming to improve the welfare of small-scale sugarcane-growers must be based on a profound knowledge of the diversity of their income-generating activities, as well as their circumstances and environments. Small-scale sugarcane farmers tend to participate in multiple activities, which can be classified into four categories: sugarcane cultivation, non-sugarcane farming activities (non-sugarcane crop production and livestock keeping), non-farm business activities (self-employment and temporary migration) and wage employment (farm wage employment and non-farm wage employment). Some research has been conducted on small-scale sugarcane contract farmers (e.g. Msuya and Ashimogo, 2005; Fernandez and Nuthall, 2009; Wendimu *et al.*, 2016; Wendimu *et al.*, 2017), but little attention has been paid to exploring how they diversify their incomes based on economic concepts and methods. Similarly, the role of CF in income diversification among small-scale farmers has received little attention with the exception of Sharma and Singh (2013), who used time-series data for the period 1970/71 to 2008/09 to examine the role of CF in crop diversification in Punjab, India. In addressing such deficiencies, we organise this chapter as follows: the subsequent sections provide respectively a literature review and present the theoretical background, followed by a description of the methodology and a presentation and discussion of the results. The final section concludes the chapter.

Literature review and theory

Most studies that focus on income diversification in the rural areas of developing countries, including Tanzania, have considered various explanatory factors for participation in certain livelihood activities (e.g. Ellis and Mdoe, 2003; Perez *et al.*, 2015; Weldegebriel *et al.*, 2015; Demissie and Legesse, 2013). The drivers behind rural farmers' participation in income-generating activities can theoretically be distinguished as 'push' and 'pull' factors. The 'pull' factors include new income opportunities emerging from market development and benefits arising from complementarities between activities, improvements to infrastructure, asset accumulation and returns from the agricultural sector (Davis and Pearce, 2001; Jalan and Ravallion, 1998; Reardon, 1997).

The 'push' factors, on the other hand, include a lack of risk-management and risk-coping strategies, a low productivity of agricultural production, insufficient liquidity and credit market failure, high transaction costs, the seasonality of agricultural activities, long-term, regular or seasonal shocks such as economic downturns, high prices of agricultural inputs, low prices of agricultural products, large fluctuations in agricultural production, droughts and floods, and a lack of assets (Ellis and Mdoe, 2003; Escobal, 2001; Reardon, 1997). Other authors include reasons for diversification born out of desperation as 'push' factors. These include poverty, the handling of livelihood risks and fluctuations in farm production due to drought, a lack of land for farming or poor productivity (Barrett *et al.*, 2001a; Reardon *et al.*, 2006).

Based on empirical and theoretical[1] reviews (e.g. Barrett *et al.*, 2001b; Corral and Reardon, 2001; Demissie and Legesse, 2013; Shi *et al.*, 2007; Lanjouw *et al.*, 2007; Winters *et al.*, 2002; de Janvry *et al.*, 2005; Escobal, 2001; Lie *et al.*, 2012; Möllers and Buchenrieder, 2005; Kimsun and Sokcheng, 2013), the factors that influence farmers' diversification behaviour in rural areas of developing countries include the following: (i) human assets (e.g. household size, age, gender and education); (ii) natural assets (e.g. land); (iii) physical assets (e.g. farming implements, durable assets); (iv) social assets (e.g. membership in associations or networks); (v) financial assets (e.g. access to formal and informal credit markets); (vi) location (e.g. distance from markets and towns); and (vii) risk (e.g. exposure to shocks).

Conceptually, asset endowments and CF variables (e.g. satisfaction with the price of sugarcane and the payment modalities practised in sugarcane CF) are important in shaping the diversification behaviour of sugarcane contract farmers in Kilombero and Turiani. The price of the contracted crop plays two roles. First, the output price affects the (relative) profitability of the contracted crop compared to other income-generating activities, thus possibly leading to an intensification of this crop or diversification into other farm and non-farm activities. Second, the output price affects household income and thus the potential for accumulating financial and non-financial assets; the availability of these assets can affect investment opportunities (e.g. procurement of physical assets such as agricultural inputs and irrigation tools) and credit constraints and thus ultimately the household's choice of livelihood strategy (e.g. intensification of the contracted crop or diversification into other activities). Household assets include natural, physical, human, and financial and social assets (see above), and they determine the ability of sugarcane farmers to participate in various farm and non-farm income-generating activities.

Methodology

Study context and data collection

The analysis presented in this chapter is based on a cross-sectional data set collected in two districts, namely Kilombero and Turiani, in July and August

2012. Both districts are dominated by farming-related activities as their main source of livelihood, where sugarcane is a dominant cash crop. Crops like maize, rice and cassava are considered to be the most important food crops, which are grown for subsistence and for selling small amounts of surplus on the market. Kilombero and Turiani demonstrate similarities in terms of land access, land fertility, market access and wage employment. Sugarcane CF is practised in both districts, although these districts differ in their implementation of the CF scheme.

The data were obtained from a cross-sectional household survey; 386 households were selected from 24,937 registered contract farmers, 8,894 in Turiani and 16,043 in Kilombero, by cluster sampling. The survey included 191 and 195 respondents from Kilombero and Turiani districts respectively. The information collected includes household composition and structure, institutional structure, crop production and other income-generating activities, as well as asset endowments. While several surveys throughout the year might give more precise estimates of income, this was not feasible within the POLICOFA project of which this survey was a part. However, the enumerators were trained in interviewing techniques and interview administration to ensure that the questions captured the targeted information. For example, in determining a household's yearly total income,[2] respondents were first asked about the activities in which they participated and then about the output obtained in local measurements and the values of these outputs using that year's price. A similar approach was applied to obtain information about the costs associated with these activities.

Empirical strategy

In the study areas, not all sugarcane-growers participated in all activities. As a result, income from various activities included a substantial number of zero values. Therefore, to examine the factors that influence sugarcane-growers' decisions to diversify, the Tobit model was used, which takes into account the fact that the dependent variable is left-censored at zero and thus avoids biased estimates (Tobin, 1958; Greene, 2003).

Our analysis adopts the empirical strategy used by Schwarze and Zeller (2005) and specifies total income from an activity as a dependent variable that quantifies the level of participation in the activity. Other studies have analysed the relative dependence on the activity by using the share of income from a specific activity in total household income or the share of time spent by the household on a specific activity in total time spent on income-generating activities as a dependent variable (e.g. Fisher, 2004; Nghiem, 2010). In addition, we use the Shannon diversification index (H_i) to investigate overall income diversification and its determinants.

To investigate the determinants of small-scale sugarcane farmers' income generation and income diversification, we use two models of two equations each. The first model relates the annual level of income from a specific category of income-generating activities to the household's capacity variables (human, physical, financial and natural capital) and incentive variables (satisfaction with the sugarcane CF arrangement). This is specified as:

$$y_{ji}^* = X_i\beta_j + u_{ji},\ j = 1,\ldots,4,\ i = 1,\ldots,N, \tag{1}$$

where:

$$y_{ji} = \begin{cases} y_{ji}^* & \text{if } y_{ji}^* > 0 \\ 0 & \text{if } y_{ji}^* \leq 0 \end{cases} \tag{2}$$

The second model relates the Shannon diversification index (H_i) to the same capacity variables and incentive variables as the model for the annual levels of income in equation (1) above. The generic model is specified as:

$$H_i^* = X_i\gamma + \varepsilon_i,\ i = 1,\ldots,N, \tag{3}$$

where:

$$H_i = \begin{cases} H_i^* & \text{if } H_i^* > 0 \\ 0 & \text{if } H_i^* \leq 0 \end{cases} \tag{4}$$

In these model specifications, N is the number of observations, y_{ji} indicates the ith household's income from the j's category of income-generating activities, y_{ji} is a partially latent variable that indicates the ith household's propensity to earn income from the j's category of income-generating activities, H_i indicates the ith household's Shannon diversification index, H_i^* is a partially latent variable that indicates the ith household's propensity to diversify its income sources, X_i is a vector of explanatory variables, β_j; $j = 1, \ldots, 4$ and γ are vectors of unknown coefficients to be estimated, and u_{ji}; $j = 1, \ldots, 4$ and ε_i are independent normally distributed error terms that are assumed to have a zero mean and a constant variance.

The models were estimated using the maximum likelihood method, which has the desirable property of being both consistent and asymptotically efficient (Tobin, 1958; Greene, 2003).

Dependent variables

In our empirical analysis, we use two types of dependent variables: (i) income from a specific category of income-generating activities (sugarcane cultivation, non-sugarcane farming activities, non-farm businesses, wage employment); and (ii) the Shannon diversification index (H_i), which measures income diversification among the four categories of income-generating activities listed above. The income is measured in monetary units and is obtained by subtracting the total costs of production from the total values of output.

The total costs of production include the costs of hired labourers and rented capital assets, but exclude the opportunity costs of (unpaid) household labour and the use of assets owned by the household. Income from sugarcane sales was obtained from the sugarcane-grower's payment invoice. The invoice indicates sales revenue, the costs of all sugarcane production activities conducted by the contractor (i.e. harvesting, loading and hauling) and other deductions like membership contribution for the year in question. Income from non-sugarcane farming activities includes income from crop and livestock production, whereas the value of production equals the value of sales earned plus the value of production consumed by the household over the 12 months preceding the survey (valued at 2012 domestic market prices for crops, animals and animal products). Income from non-farm businesses is the sum of all amounts earned by a smallholder sugarcane-grower from self-employment activities and temporary migration over the previous 12 months. Income from wage employment includes income from jobs within and outside the agricultural sector. Subsequently, total household income from all sources was computed by adding all net incomes from sugarcane cultivation, non-sugarcane farming, non-farm businesses and wage employment. The Shannon diversification index (H_i) was calculated as follows:[3]

$$H_i = -\sum_{j=1}^{M_i} s_{ij} \log(s_{ij}),\ i = 1,\ldots,N, \tag{5}$$

where M_i represents the number of diversified income sources of household i and s_{ij}; $j=1,\ldots,\ M_i$ indicates the share of income from activity j in total household income. The Shannon diversification index (H_i) takes into account the number of income sources and their evenness. The index is zero for households with only one single source of income and increases with higher income diversification.

Explanatory variables

Conceptually, the key explanatory variables that are assumed to influence sugarcane-growers' decisions to diversify their income-generating activities include asset endowments (physical, social, financial, human and natural) and the farmer's satisfaction with the sugarcane CF arrangement. Regarding physical assets, area of land owned, measured in acres, is used. Access to physical assets is fundamental for participation in many income-generating activities, and hence it ultimately influences income diversification. For example, access to land may make it easier for sugarcane smallholders to participate in various agricultural activities. However, through increased income from agricultural activities, these farmers may be able to invest more in non-sugarcane farming and non-farm business activities. Conversely, increased access to land may have a negative influence on income from non-farm businesses and wage employment because sugarcane smallholders may opt to assign more labour to agricultural activities.

To capture human assets, three variables were used, namely education and age of the household head, and number of household members of working age (16–65 years). This study anticipates that the age of the household head, measured in years, is positively correlated with income from different sources. This is because older household heads are likely to have more experience and stronger networks that are relevant for farming activities, non-farm businesses and wage employment opportunities. Similarly, the number of household members of working age is predicted to have a positive relationship with diversification. A household with a large workforce (number of household members of working age) is likely to have a higher probability of participating in different income-generating activities due to the ability to supply more labour (e.g. Reardon, 1997).

Satisfaction with the CF agreement is a further explanatory variable. This is a dummy variable that has a value of one if a smallholder sugarcane-grower is satisfied with the pre-determined price of sugarcane and the payment modalities agreed in the contract and is zero otherwise.

Access to physical infrastructure (e.g. paved roads) usually creates better opportunities for smallholders to diversify. Access to roads lowers transportation costs and other transaction costs and hence may increase smallholders' opportunities to increase and diversify their incomes. Increased incomes may lead to further income diversification through the accumulation of capital for further businesses.

Results and discussion

Descriptive statistics

Table 9.1 presents descriptive statistics of the total income and the dependent variables and explanatory variables in our empirical analysis. Average income from sugarcane cultivation is significantly higher in Kilombero than in Turiani, while sugarcane-growers in Turiani have on average a significantly higher income from non-sugarcane farming activities than households in Kilombero (1 per cent level of significance). Concerning land ownership, the average size of total farm land was 5.31 acres in Kilombero and 8.08 acres in Turiani. In terms of education, household heads had on average 6.3 years of education in Kilombero and 6.0 years in Turiani. This finding suggests that the majority of smallholders in the two study areas had less than primary school education, which may restrict them in diversifying into well-paid non-agricultural activities. This is consistent with the findings of Tien (2014), namely that most workers with lower levels of education participate in low-paid jobs in Vietnam. The average age of household heads was 46 in Kilombero and 53 in Turiani. About half (49 per cent) of the sugarcane-growers in Kilombero were satisfied with their CF arrangements, but only 8 per cent in Turiani. Dissatisfaction with CF arrangements, particularly in Turiani, could be a major push factor towards income diversification.

Table 9.1 Summary of descriptive statistics

Variable	*Description*	*Kilombero*	*Turiani*	*Overall*
Dependent variables				
Total income (000' Tsh)		3,247.4	2,627.5	2,934.2
Income from sugarcane cultivation (000' Tsh)		1,548.5	506.7	1,022.2
Income from non-sugarcane farming (000' Tsh)		584.8	1,013.5	801.4
Income from non-farm businesses (000' Tsh)		958.3	876.4	916.9
Income from wage employment (000' Tsh)		155.7	231.8	194.1
Shannon Index (H)	Shannon diversification index	0.31	0.35	0.33
Explanatory variables				
Land owned	Total farm holdings in acres	5.31	8.03	6.71
Education	Household head's years of schooling	6.31	6.02	6.16
Age	Age of the household head in years	45.95	53.21	49.62
Workforce	Number of household members in working age	2.34	2.13	2.26
Infrastructure	Access to road (0/1)	0.05	0.04	0.05
CF satisfaction	Sugarcane grower's satisfaction with price and payment modalities (0/1)	0.49	0.03	0.28
Location (Kilombero)	Household residing in Kilombero (1) or Turiani (0)	1	0	0.49
Number of observations		191	195	386

Source: authors' fieldwork (2012).

Activity participation rates and shares in total income

Table 9.2 presents descriptive statistics regarding participation rates and income shares for the four different categories of income-generating activities. This table indicates that the sugarcane-growers in the study area are participating in multiple income-generating activities. Many of them are involved in non-sugarcane farming activities, particularly in Turiani. About a quarter of the sugarcane farmers are involved in non-farm businesses, while only few of them earn an income from wage employment (7 per cent). Sugarcane cultivation is the most important income-generating activity in Kilombero. The average income from sugarcane cultivation represents 48 per cent of the average total household income, and the average income share among all the households in Kilombero is even 65 per cent. In contrast, in Turiani, non-sugarcane farming is the most important income-generating activity. The average income from non-sugarcane farming represents 39 per cent of the average total household income in Turiani, and the average

Table 9.2 Activity participation rates and shares in total income*

Activity	*Kilombero (N=191)*	*Turiani (N=195)*	*Both areas (N= 386)*
Sugarcane cultivation			
Participation rate (%)	100	100	100
Proportion with positive income (%)	94	81	88
Mean income (Tsh)	1,548,530	506,721	1,022,228
Share of mean income (%)	48	19	35
Mean income share (%)	65	24	45
Non-sugarcane farming			
Participation rate (%)	40	87	64
Mean income (Tsh)	584,827	1,013,532	801,401
Share of mean income (%)	18	39	27
Mean income share (%)	19	56	38
Non-farm businesses			
Participation rate (%)	19	27	23
Mean income (Tsh)	958,314	876,356	916,911
Share of mean income (%)	30	33	31
Mean income share (%)	12	14	13
Wage employment			
Participation rate (%)	5	8	7
Mean income (Tsh)	155,712	231,754	194,127
Share of mean income (%)	5	9	7
Mean income share (%)	3	5	4

Source: authors' fieldwork (2012).

Notes

* 16 observations with negative total income (7 in Kilombero and 9 in Turiani) were excluded from the calculations of the mean income shares.

income share among all surveyed households is even 56 per cent in Turiani, while the corresponding figures for income from sugarcane cultivation are only 19 per cent and 24 per cent, respectively. In line with this, sugarcane farmers in Kilombero earn on average much more from sugarcane cultivation than sugarcane farmers in Turiani. The average income from sugarcane in Kilombero is more than three times the average in Turiani. In contrast, households in Turiani on average have almost twice the income from non-sugarcane farming compared to households in Kilombero. The differences in average non-sugarcane income (i.e. non-sugarcane farming income, non-farm business income, plus income from wage employment) between Kilombero and Turiani are statistically significant at 1 per cent significance level.

Figure 9.1 presents income shares for households with different sizes of land ownership. This indicates that, in both study locations, diversification generally decreases with increases in land ownership. The figure also shows that the share of sugarcane income monotonically increases with agricultural land holdings in Kilombero, while in Turiani the share is largest for households with medium-sized agricultural land holdings (1–4 ha). In Turiani, the share of income from non-sugarcane farming significantly increases with agricultural land ownership.

In Kilombero, sugarcane cultivation is the most important source of income for households with more than 1 ha of land, while for households with less than 1 ha of land, income from non-farm businesses has about the same importance

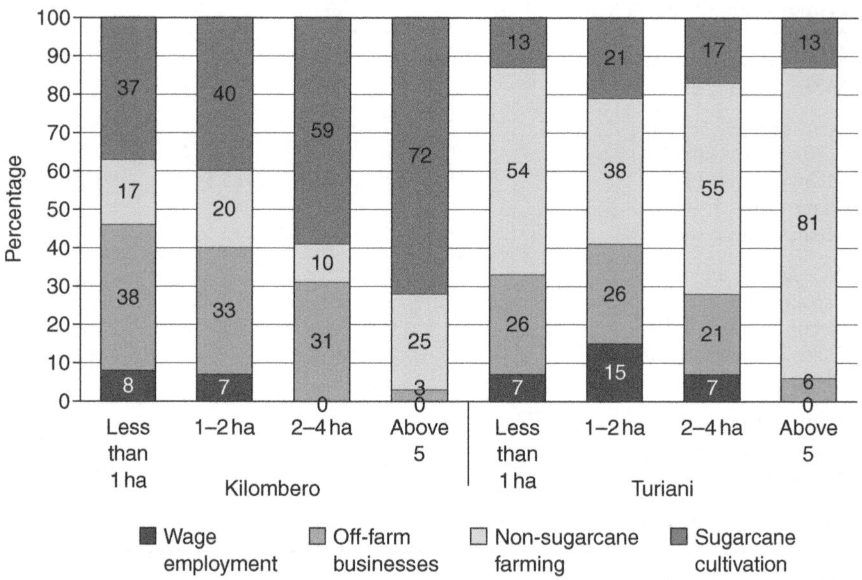

Figure 9.1 Income shares by size of land ownership in the two study areas.

Source: authors' fieldwork (2012).

as income from sugarcane cultivation. Meanwhile, in both study regions, the importance of income from non-farm businesses decreases with increasing land ownership. In Turiani, non-sugarcane farming activities are the most important source of income for all categories of landholding. This is in line with a study by Nkonya and Barreiro-Hurle (2012), who argue that sugarcane cultivation is no longer a priority activity in the Turiani area as it used to be with the out-grower system that was terminated following the privatisation of the sector.

Determinants of income diversification

The estimation results of the Tobit models that analyse sugarcane-growers' incomes from the four categories of income-generating activities are presented in Tables 9.3 and 9.4. Table 9.3 presents the estimation results of the model for sugarcane farming income. We conducted separate estimations for the two study locations (Kilombero and Turiani) because a Chow test indicates that the slope coefficients of this model differ significantly between the two study areas (P-value 0.028, when allowing the intercept to differ between the two study locations), so that a pooled model for the two study areas seems to be inappropriate. Table 9.4 presents the estimation results of the models for the other three sources of income, those for non-sugarcane farming income, non-farm business income and wage employment income, respectively. These models are estimated using the pooled data set for both study locations (Kilombero and Turiani) but allow the intercept to differ between

Table 9.3 Tobit maximum likelihood estimates of the determinants of income from sugarcane cultivation

Variable	*Income from sugarcane cultivation*			
	Kilombero (N = 191)		*Turiani (N = 195)*	
	Coefficient	*t-ratio*	*Coefficient*	*t-ratio*
Constant	1,074,733	1.09	−117,642	−0.31
Land owned	59,893	5.19***	15,062	2.15**
Education	49,025	0.70	37,774	1.36
Age	−3,912	−0.30	4,325	0.91
Workforce	119,283	0.91	−34,945	−0.62
Infrastructure	−184,527	−0.24	600,727	1.71*
CF satisfaction	−602,225	−1.69*	250,370	0.99
Left-censored	11		37	
Uncensored	180		158	
Log-likelihood	−2,898.95		−2,431.69	
L.R. $chi^2(6)$	33.28***		11.12*	

Source: authors' fieldwork (2012).

Notes
***, ** and * imply statistical significance at 1%, 5% and 10% levels, respectively.

Table 9.4 Tobit maximum likelihood estimates of the determinants of income from non-sugarcane farming, non-farm businesses and wage employment

Variable	*Non-sugarcane farming income*		*Non-farm businesses income*		*Wage employment income*	
	Coefficient	*t-ratio*	*Coefficient*	*t-ratio*	*Coefficient*	*t-ratio*
Constant	811,396	1.70*	–2,554,202	–0.76	–6,256,991	–1.86*
Land owned	40,407	6.27***	27,620	0.60	–141,521	–1.09
Education	–4,788	–0.14	576,025	2.12**	791,835	2.49**
Age	–9,205	–1.51	–179,432	–3.78***	–122,922	–2.45**
Workforce	143,858	2.18**	755,040	1.66*	215,989	0.50
Infrastructure	54,310	0.14	–1,310,478	–0.47	4,515,702	2.41**
CF satisfaction	–636,192	–2.82***	–50,961	–0.03	–849,357	–0.55
Location (Kilombero)	–844,830	–4.26***	–2,403,690	–1.80*	–2,736,346	–2.05**
Left-censored	140		298		360	
Uncensored	246		88		26	
Log-likelihood	–3,945.26		–1,630.63		–490.11	
L.R. $chi^2(7)$	96.85***		30.97***		32.81***	

Source: authors' fieldwork (2012).

Notes
***, ** and * imply statistical significance at 1%, 5% and 10% levels, respectively.

the two study locations. A Chow test indicates that the slope coefficients of the models for incomes from non-sugar farming activities and from non-farm businesses do not significantly differ between the two study locations (P-values 0.55 and 0.40, respectively). Due to low participation rates in wage employment (10 households in Kilombero and 16 households in Turiani), the model for income from wage employment could not be separately estimated for the two study locations due to non-convergence, so we present the results based on the pooled data set for both study locations.

As expected, the results indicate that the area of agricultural land owned by the household significantly increases income from farming activities (both sugarcane cultivation and non-sugarcane farming), that is, sugarcane-growers with more agricultural land tend to earn more income from farming than households that own little land. While the agricultural land area owned by the household has about the same effect on income from non-sugarcane farming in the two study locations (23,603 Tsh/acre), its effect on income from sugarcane cultivation is much higher in Kilombero (44,711 Tsh/acre) than in Turiani (10,131 Tsh/acre).[4] This finding is in line with Figure 9.1, which shows – although not in a *ceteris paribus* context – that households in Kilombero focus particularly on sugarcane cultivation with increasing land ownership, while this is not the case for households in Turiani.

In contrast to the effects on income from agricultural activities, the area of agricultural land owned by the household has no significant effect on income from non-farm activities (non-farm businesses and wage employment). This indicates that the potential positive effects of land ownership on non-farm income (e.g. more resources for investments in these activities) and the potential negative effects (e.g. less household labour available for these activities) cancel out.

On the other hand, while the education and age of the household head have no significant influence on income from farming activities (sugarcane cultivation and non-sugarcane farming), they significantly affect income from non-farming activities (both non-farm businesses and wage employment) (Tables 9.3 and 9.4). As expected, education increases income from both non-farm businesses (109,011 Tsh/year of education) and wage employment (27,416 Tsh/year of education). These findings are consistent with the results of Rahut and Micevska Scharf (2012), who found that education has a positive effect on overall non-farm household income in the Himalayas, and they are partly consistent with the results of Demissie and Legesse (2013), who found that education had a positive effect on income from non-farm self-employment, but a negative effect on income from non-farm wage employment in Ethiopia. In contrast to our expectations, the age of the household head has a negative effect on income from non-farm businesses and wage employment. This finding is in line with Demissie and Legesse (2013), who found that the age of the household head had a negative effect on income from both non-farm self-employment and non-farm wage employment.

Our estimation results also show that the number of household members of working age has a positive and significant influence on income from non-sugarcane farming activities and non-farm business income. This implies that smallholder sugarcane-growers who have many members of working age were able to participate in different agricultural and non-farm business activities and to earn more income from these activities than those who only have a small number of household members of working age. This is consistent with the study by Demissie and Legesse (2013). Likewise, access to infrastructure such as roads has a positive and significant influence on income earned from wage employment, a finding that corroborates that of Escobal (2001).

Farmers' levels of satisfaction with their sugarcane CF arrangements are negatively associated with sugarcane income in Kilombero and with non-sugarcane farming income in both study locations. This means that households that are satisfied with their sugarcane CF arrangements earn *ceteris paribus* less income from sugarcane cultivation in Kilombero and from non-sugarcane farming in both study locations than households that are dissatisfied. Our results also indicate that households in Kilombero have *ceteris paribus* a lower income from non-sugarcane farming activities, non-farm businesses and wage employment activities than households in Turiani.

Determinants of overall income diversification

Table 9.5 presents the results of the Tobit model for analysing the determinants of overall income diversification. A Chow test indicates that the coefficients (except for the intercept) do not significantly differ between the two

Table 9.5 Tobit results for the determinants of overall diversification

Variable	*Coeff.*	*t-ratio*
Constant	0.427	3.08***
Land owned	0.004	2.04**
Education	0.010	1.07
Age	–0.003	–1.91*
Workforce	0.028	1.46
Infrastructure	0.385	3.44***
CF satisfaction	–0.052	–0.83
Location (Kilombero)	–0.267	–4.74***
Left censored	136	
Uncensored	250	
Log-likelihood	–273.89	
L.R. $chi^2(7)$	54.67***	

Source: authors' fieldwork (2012).

Notes
***, **, and * indicate significance levels of 1%, 5%, and 10%, respectively.

study locations (P-value 0.94). Hence, we present the estimation results based on the pooled data set. The results reveal that the area of land owned has a positive influence on income diversification, that is, as land area increases, households have, *ceteris paribus*, a more diversified income.

Access to paved roads has a positive impact on income diversification, as expected. While Schwarze and Zeller (2005) and Sharma and Singh (2013) find that households with a higher educational level had a larger number of sources of income than households with a lower education level, we did not find that the level of education affects income diversification in our study.

Our results also show that households in Kilombero have, *ceteris paribus*, a less diversified income than households in Turiani. The age of the household head has a negative and marginally statistically significant (P-value 0.057) influence on income diversification. This indicates that sugarcane-growing households with older household heads have, *ceteris paribus*, a less diversified income than households with younger household heads.

Concluding remarks and policy implications

The objective of this chapter has been to determine the factors that influence income diversification among sugarcane-growers. For rural communities, income diversification is a key livelihood option (Ellis, 1998). A good understanding of the factors that influence rural households' incomes from various income-generating activities is essential for designing informed strategies that focus on rural development.

Sugarcane-growers' non-sugarcane income-generating activities with the highest participation rates are non-sugarcane farming activities (40 per cent in Kilombero, 87 per cent in Turiani), followed by non-farm businesses (19 per cent in Kilombero, 27 per cent in Turiani) and finally wage employment (5 per cent in Kilombero, 8 per cent in Turiani). Thus, households in Turiani have higher participation rates in all non-sugarcane income-generating activities than households in Kilombero. However, average total household income is higher for households in Kilombero than for their counterparts in Turiani.

Our econometric analysis indicates that the key factors that influence sugarcane-growers' incomes from four different activities include the area of land owned by the household, the education and age of the household head, the household's workforce, access to paved roads, satisfaction with the contract farming arrangement and location. Regarding land ownership, the results indicate that sugarcane-growers who have a larger endowment of land have a higher farm income (sugarcane cultivation and other farming activities). As expected, the results for the household's workforce indicate a positive and significant influence on income from non-sugarcane farming activities and non-farm business income, indicating that sugarcane smallholders with more household members of working age are able to earn more income from these two activities than households with fewer family members of working age.

Sugarcane-growing households in Kilombero were found to have, *ceteris paribus*, a less diversified income than sugarcane-growing households in Turiani. Households in Kilombero earned less income from non-sugarcane farming, non-farm businesses and wage employment, but earned more income from sugarcane cultivation than households in Turiani. As access to roads has a positive effect on income diversification, infrastructural development such as paved roads may facilitate smallholder farmer's income diversification. However, as the paved roads are usually the main (most frequently used) roads currently, while the unpaved roads are the less important (less frequently used) roads, the estimated effect of being located on a paved road includes both the effect of being located on a main road and the effect of the road being paved. Hence, the effect of paving further roads will probably be smaller than indicated by our estimation results because paving an additional road will not necessarily make this road a main (most frequently used) road.

Reducing poverty among the rural population is one of the main concerns of Tanzania's Developing Vision 2025. Sugarcane cultivation is recognised as one of the potential crops for poverty alleviation in rural areas of Tanzania. However, as income diversification usually mitigates income risk, poor farm households with limited savings may be better able to cope with income risk if they diversify their incomes to a certain extent rather than rely too much on a single income-generating activity such as sugarcane cultivation. Furthermore, not only farming activities but also non-farming activities could play an important role in the development of rural villages. As we found significantly positive effects of education on income from non-farm activities (non-farm businesses and wage employment), and as participation in the education system has increased over time (resulting, for example, in a negative correlation between age and education years in our data set), increased participation in the education system will likely contribute to increased non-farm income in rural villages. Our results also indicate that households with younger household heads have more income from non-farm activities than households with older household heads. This finding could contribute to increasing average incomes of rural households from non-farm activities in the future if households maintain their income levels, even though their household heads grow older every year, and the replacement of old household heads by young household heads increases the non-farm incomes of these households. New research based on longitudinal data could investigate whether this is indeed the case or whether households' non-farm incomes decline with the increasing age of their household heads.

Notes

1 Various theoretical models have been used to analyse income diversification with a focus on non-farm income by using the rational choice approach, which focuses on monetary incentives.

2 Total income is the sum of the net income from all the household's income-generating activities. The net income was calculated by subtracting the total costs of production from the total values of production, including subsistence products consumed by the household.

3 The Shannon diversification index is not defined for negative (income) shares, but a few households in our sample have a negative income from sugarcane production. In order to be able to calculate the income diversification for all the households in our sample, we set the negative sugarcane incomes to zero before calculating total income, income shares and the Shannon diversification index of income.

4 The selected marginal effects given in the text indicate the marginal effects on the unconditional expected value of the dependent variable at the average values of the explanatory variables. Due to space limitations, tables giving the marginal effects of all explanatory variables on all dependent variables are not given here but are available from the authors upon request.

References

Barrett, C. B., Bezuneh, M. and Abdillahi, A. (2001a). Income Diversification, Poverty Traps and Policy Shocks in Côte d'Ivoire and Kenya. *Food Policy*, 26(4), 367–384.

Barrett, C. B., Reardon, T. and Webb, P. (2001b). Nonfarm Income Diversification and Household Livelihood Strategies in Rural Africa: Concepts, Dynamics, and Policy Implications. *Food Policy*, 26(4), 315–331.

Corral, L. and Reardon, T. (2001). Rural Nonfarm Incomes in Nicaragua. *World Development*, 29, 427–442.

Davis, J. and Pearce, D. (2001). The Rural Nonfarm Economy in Central and Eastern Europe. In Z. Lerman and C. Csaki (eds), *The Challenge of Rural Development in the EU Accession Process*. Washington, DC: World Bank.

De Janvry, A., Sadoulet, E. and Zhu, N. (2005). *The Role of Nonfarm Income in Reducing Poverty and Inequality in China*. CUDARE Working Paper No. 1001. University of California, Berkeley.

Demissie, A. and Legesse, B. (2013). Determinants of Income Diversification among Rural Households: The Case of Smallholder Farmers in Fedis District, Eastern Hararghe Zone, Ethiopia. *Journal of Development and Agricultural Economics*, 5(3), 120–128.

Ellis, F. (1998). Survey Article: Household Strategies and Rural Household Diversification in Developing Countries. *Journal of Agricultural Economics*, 51(2), 289–301.

Ellis, F. (2003). A Livelihoods Approach to Migration and Poverty Reduction. Paper commissioned by the Department for International Development (DFID). Contract No. CNTTR 03 4890. Online: www.uea.ac.uk/documents/439774/621632/F+Ellis+2003+Livelihoods+Migration.pdf. Accessed 19 September 2016.

Ellis, F. and Mdoe, N. (2003). Livelihoods and Rural Poverty Reduction in Tanzania. *World Development*, 31(8), 1367–1384.

Escobal, J. (2001). The Determinants of Nonfarm Income Diversification in Rural Peru. *World Development*, 29(3), 497–508.

Fernandez, M. D. P. and Nuthall, P. L. (2009). Technical Efficiency in the Production of Sugarcane in Central Negros Area, Philippines: An Application of Data Envelopment Analysis. *ISSAAS*, 15(1), 77–90.

Fisher, M. (2004) Household Welfare and Forest Dependence in Southern Malawi. *Environment and Development Economics*, 9(2), 135–154.

Greene, W. (2003). *Econometric Analysis* (5th edn). Englewood Cliffs, NJ: Prentice Hall.

Jalan, J. and Ravallion, M. (1998). *Geographic Poverty Traps?* Discussion Paper 86, Institute for Economic Development, Boston University.

Kimsun, T. and Sokcheng, P. (2013). *The Role of Income Diversification during the Global Financial Crisis: Evidence from Nine Villages in Cambodia.* CDRI Working Paper Series No. 78, Cambodia Development Resource Institute, Phnom Penh, Cambodia.

Lanjouw, P., Shariff, A. and Rahut, D. B. (2007). *Rural Non-farm Employment in India: Access, Income, Farm, Poverty Impact.* Working Paper No. 81, National Council of Applied Economic Research (NCAER), New Delhi, India.

Lie, H., Rich, K. M., Kurwijila, L. R. and Jervell, A. M. (2012). Improving Small-holder Livelihoods through Local Value Chain Development: A Case Study of Goat Milk Yogurt in Tanzania. *International Food and Agribusiness Management Review*, 15(3), 55–86.

Little, P. D. and Watts, M. J. (1994). *Living under Contract: Contract Farming and Agrarian Transformation in Sub-Saharan Africa.* Madison: University of Wisconsin Press.

Möllers, J. and Buchenrieder, G. (2005). Theoretical Concepts for the Analysis of Non-farm Rural Employment. *Quarterly Journal of International Agriculture*, 44(1), 19–36.

Msuya, E. E. and Ashimogo, G. (2005). *Estimation of Technical Efficiency in Tanzania Sugarcane Production: A Case Study of Mtibwa Sugar Estate Outgrowers Scheme.* MPRA Paper No. 374. Online: http://mpra.ub.uni-muenchen.de/3747. Accessed 22 September 2016.

Nghiem, L. T. (2010). Activity and Income Diversification: Trends, Determinants and Effects on Poverty Reduction – The Case of the Mekong River Delta. PhD Thesis. Erasmus University, Rotterdam; Shaker Publishing BV, Maastricht.

Nkonya, N. and Barreiro-Hurle, J. (2012). *Analysis of Incentives and Disincentives for Sugar in the United Republic of Tanzania.* Technical Notes Series, Monitoring and Analysing Food and Agricultural Policies (MAFAP), FAO, Rome.

Perez, C. V., Bilsborrow, R. and Torres, B. (2015). Income Diversification of Migrants Colonists vs Indigenous Populations: Contrasting Strategies in the Amazon. *Journal of Rural Studies*, 42, 1–10.

Rahut, D. B. and Micevska Scharf, M. (2012). Livelihood Diversification Strategies in the Himalayas. *Australian Journal of Agriculture and Resource Economics*, 56, 558–582.

Reardon, T. (1997). Using Evidence of Household Income Diversification to Inform Study of the Rural Nonfarm Labor Market in Africa. *World Development*, 25(5), 735–747.

Reardon, T., Berdegue, J., Barrett, C. B. and Stamoulis, K. (2006). Household Income Diversification into Rural Non-farm Activities. In S. Haggblade, P. Hazell, and T. Reardon (eds), *Transforming the Rural Nonfarm Economy* (pp. 115–140). Baltimore, MD: Johns Hopkins University Press.

SBT (2007). *Sugar Cane Supply Agreement between Mtibwa Sugar Estate and Mtibwa Out-growers Association.* Report from Sugar Board of Tanzania.

SBT (2014). *Production Data: Sugar Production for 2002/03–2013/14.* Sugar Board of Tanzania. Online: www.sbt.go.tz/index.php/production-datas. Accessed 26 September 2016.

Schwarze, S. and Zeller, M. (2005). Income Diversification of Rural Households in Central Sulawesi, Indonesia. *Quarterly Journal of International Agriculture*, 44(1), 61–73.

Sharma, N. and Singh, S. P. (2013). Agricultural Diversification and Contract Farming in Punjab. *Journal of Economic and Social Development*, 9(1), 38–51.

Shi, X., Heerink, N. and Qu, F. (2007). Choices between Different Off Farm Employment Sub-categories: An Empirical Analysis for Jiangxi Province, China. *China Economic Review*, 18, 438–455.

Tien, N. D. L. (2014). *An Analysis of Labour Market Returns to Education in Vietnam: Evidence from the National Labour Force Survey 2012*. Turin School of Development Working Paper No. 3, International Training Centre of the ILO, Turin, Italy.

Tobin, J. (1958). Estimation of Relationship for Limited Dependent Variables. *Econometrica*, 26(1), 24–36.

Weldegebriel, Z. B., Follon, G. and Prowse, M. (2015). The Determinants of Non-farm Income Diversification in Rural Ethiopia. *Journal of Poverty Alleviation and International Development*, 6(1), 109–130.

Wendimu, M. A., Henningsen, A. and Czekaj, T. G. (2017). Incentives and Moral Hazard: Plot Level Productivity of Factory-Operated and Outgrower-Operated Sugarcane Production in Ethiopia. *Agricultural Economics*, forthcoming, DOI:10.1111/agec.12356.

Wendimu, M. A., Henningsen, A. and Gibbon, P. (2016). Sugarcane Outgrowers in Ethiopia: 'Forced' to Remain Poor? *World Development*, 83, 84–97.

Winters, P., Davis, B. and Corral, L. (2002). Assets, Activities and Income Generation in Rural Mexico: Factoring in Social and Public Capital. *Agricultural Economics*, 27(2), 139–156.

World Bank (2008). *World Development Report 2008: Agriculture for Development*. Washington, DC: World Bank.

World Bank (2011). *Making the Grade: Smallholder Farmers, Emerging Standards and Development Assistance Programs in Africa*. A Research Program Synthesis No. 62324 – AFR. Washington, DC: World Bank.

Part IV

Alternative aspects of contract farming

10 Trusting your partner?

Sunflower contract farming in central Tanzania

Frederik Brønd

Introduction

Contract farming is often seen as a means to reduce the marketing risks that exist between the producers and buyers of agricultural products. Instead of only relying on trust and chance in transactions, reference can also be made to a contract to overcome the problem of a lack of trust and associated transaction costs. However, as the coordination costs for both buyer and seller associated with the formal enforcement of contracts are high in developing-country contexts, trust remains an important component in contract farming arrangements.

This is the main argument of this chapter on sunflower contract farming in central Tanzania. The chapter suggests that the production and perception of trust are important for coordination between farmers and buyers, as well as for the territoriality of the value chain, that is, the locations in which sunflower contracts unfold. Thus, the relational production of agricultural economies based on contract farming creates territorial differences in the rural context. Where the patchy inclusion of smallholders in global value chains may create 'privileged spaces' on the basis of the material properties of the territory concerned (e.g. land quality, infrastructure, labour), this chapter argues that an investigation of personal relations among and between the contracting parties locally is also important to arrive at an understanding of where and how the value chains 'touches down'. The chapter argues that the global value chain (GVC) approach looks too much at the material characteristics of the transaction in order to determine the characteristics of GVC governance and too little on the non-economic aspects of relations between buyer and seller, no matter how simple the transaction may be. Gereffi *et al.* (2005) outline a typology of five different forms of governance depending on three different characteristics: one concerning the suppliers (capabilities) and two concerning the transaction (codifiability and complexity). This chapter argues that the actual nature of the relations between supplier and buyer will affect the type of coordination and that it is insufficient to look at the material characteristics alone.

The chapter will begin by outlining how trust has been used as an analytical concept in economic sociology, institutional economics and political economy and showing how the notion of trust can be applied in the context of contract farming. Second, the context of the case study will be described. The following three sections present the analysis of the case study. First the role of trust in the contract arrangement will be explored, both in terms of how trust is produced between buyer and seller, and how it is used to sustain the relationship. Second, the chapter explores the strategies that the sourcing companies follow to deal with the lack of or the production of trust. Lastly, it is argued that the production and use of trust is geographical and may produce or sustain rural inequalities in the agricultural economy.

Contract farming and the role of trust

As Oya (2012) outlines contract farming in developing countries has been studied from a variety of perspectives since the 1980s. He identifies several 'schools' within the literature on contract farming, ranging from neo-populist and political economy perspectives to management and econometric studies, all bringing different methodologies and ideological standpoints into the research. These differences are exposed by the various schools' efforts to answer fundamentally different questions, such as 'When is contract farming economically efficient?', 'What are the (welfare) impacts of contract farming?' and 'What are the causes of contract farming from the point of view of political economy?'

As an analytical concept, however, trust has only attracted modest attention in studies on contract farming. This may be due to the fact that contract farming has traditionally been seen as a means to reduce the need for trust and personal relations by replacing them with rules and sanctions. When incorporated, trust has been integrated into the study of contract farming in relation to enforcement and compliance with contracts (e.g. Warning and Key, 2002; Narayanan, 2012). However, trust has a long history of being incorporated into studies of economic relations by economic sociologists and institutional economists. It has been studied in various ways, as a precondition for markets, an economic asset of individuals (in line with social capital) or an instrument to decrease transaction costs in complex firm-to-firm interactions. These different entry points into theoretical perspectives on trust are elaborated below.

First, trust has for a long time been studied as a precondition for markets, acknowledging that markets do not exist in a vacuum apart from the rational self-interest of economic agents. General equilibrium models, for instance, assume perfect information and thus no uncertainty – the precondition for trust. In opposition to this market fundamentalist view, Granovetter (1985) argues that the economic behaviour of individuals is embedded in social relations in which trust especially plays an important role in making market interactions happen.

Second, apart from being a precondition for a market economy (Platteau, 1994), trust also has the function of reducing the costs of a specific transaction. If two actors rely on trust to engage in a deal, such as an agreement on future delivery, the costs associated with controlling the other part, searching information, writing contracts etc. are reduced and pave the way for a more efficient economic coordination.

Third, and of particular importance to contract farming in developing countries, there has been notable discussion of the perceived lack of trust in developing economies. It is argued that this failing hinders economic specialisation and reductions in transaction costs in economic transactions (Humphrey and Schmitz, 1998), both of which reduce the potential for economic growth. Some scholars, such as Platteau (1994) and Fukuyama (1995), argue that the overall morality inherent in a society's culture plays a key role in the building of trust, thus sustaining an efficient market, and that this generalised morality is stronger in Western countries. This view is opposed by many others (e.g. Harriss, 2003; Moore, 1999), who argue for the decisive importance of social interactions in building up trust in the market place. However, there does seem to be some agreement that markets in developing countries are more dependent on personal trust, given that legal institutions are weak, thus making it more difficult to rely on macro-level sanctions (Harriss, 2003; Lyon, 2000; Humphrey and Schmitz, 1998). Humphrey and Schmitz (1998), for example, argue that impersonal macro-level trust (e.g. state institutions) and meso-level trust (e.g. kinship) have been eroded in post-colonial states, while micro-level (interpersonal) trust seems to be carrying increased weight.

It is important to note, however, that the focus on trust and 'social capital', which in Putnam's definition is understood as 'networks, norms and trust that facilitate co-operation and co-ordination' (Lyon, 2000), has been criticised for ignoring the power relations that also structure relations between individuals. In his study of trust between Indian businesses, for example, Harriss (2003) argues that 'the language of "trust" frequently conceals relationships of power like those that obtain both within and between firms'. In the same vein, Bachmann (2006) argues that trust and power often occur in combination as means of controlling relations and that, because of power's role in establishing hierarchies and institutions, this also is a precondition of trust.

The political economy tradition of research on contract farming has also produced insights that are relevant in understanding the nature of the relationship involved in a contract farming arrangement. Acknowledging the breadth of the political economy literature on contract farming, one particular concept provides an interesting perspective: 'the moral economy of the contract'. As introduced by Clapp (1994), it is not only the nature of the relationship between growers and companies that is interesting, but also how this relationship is becoming an ideological representation that reconciles smallholders to their subordination to capital. Clapp draws on empirical material from a cooperative's contract (bananas) with United Foods in Honduras and a brewery's contracts with barley-producing smallholders in Peru. In the Peruvian case, the brewery builds a social order – or in Clapp's words, a 'hegemony' – of the contract whereby it abstains from pursuing minor transgressions of the contract (the farmers use some of the fertiliser supplied by the brewery for their own potato-growing) and in this way obtains the authority to subordinate the farmers in the relationship to itself. Narayanan (2012) uses the same concept in a study of contract farming in Indian agriculture. Here, minor transgressions by both farmers and companies are forgiven *ex post*, and the relationship between them is much wider in scope than the formal contract would suggest: companies employ the family members of contracted farmers, and the companies' extension officers often provide services to farmers other than the contracted crops. From the farmers' perspective, the moral economy of the contract impels them to maintain the system or 'social order'. Where trust between economic actors is necessary to understand why contracts are signed and upheld, the notion of the moral economy of the contract provides a further dimension to this issue, thus turning the arrangement into a symbol of the political and ideological order. In this way, the production of trust in the contract arrangement can be interpreted as a means by which the hegemony is reproduced and thus as a strategy available to companies to ensure efficient coordination by reducing coordination costs.

Trust has been used before as an explanatory factor in the production of economic outcomes. Playing a similar role as institutions, trust can mediate the rolling out of economic relations, thus reducing transaction costs and providing spaces of inclusion and exclusion in situations of low and high levels of the production of trust between economic agents. In fact, using the definition of Neilson and Pritchard (2009), trust can be seen as an institution since it is a self-imposed form of conduct often (but not always) associated with some enforcement characteristic. However, Neilson and Pritchard's focus is clearly on more collectively defined institutions forged by cultural and place-based struggles, for instance, production and product standards. Therefore the concept of trust adds to the analytical dimensions provided by studies of contract farming, and when incorporated into the analysis of an agricultural value chain, it can be helpful for understanding why contract farming sometimes succeeds and sometimes fails in these specific economic and personal relationships between buyer and grower. In this chapter, then, trust will be used as a platform for understanding when coordination between sourcing companies and farmers is successful and when it fails. Furthermore, the concept of trust will be used to explain the strategic choices companies make in order to ensure efficient coordination by producing a moral economy between farmers and buyers.

Sunflower contract farming in Dodoma: the setting

The basic idea underlying the specific contract farming intervention in Dodoma (central Tanzania) that is the subject of this chapter was conceived by a local NGO called the Rural Livelihood Development Company (RLDC) funded by the Swiss development agency. Adopting a 'making markets work for the poor' (M4P) approach (RLDC, 2014), this NGO identified the lack of quality seeds for planting to be a main constraint on smallholders trying to improve their income from growing sunflowers. To increase the accessibility of seeds for farmers, RLDC introduced contract farming as a solution, thereby introducing a non-market institution to solve the market problem. Local small and medium-sized oil-milling companies were encouraged to participate in the scheme by designing and implementing the contract farming arrangements themselves, while the NGO would finance the capacity-building of farmers, as well as an HIV/AIDS intervention in the villages incorporated into the contract farming arrangements.

Data collected in 2013 consisted of interviews with company managers or owners of the eight companies in the Dodoma region and farmers in three villages (Kiegea, Chamtumile and Berege). In these villages, two of the eight companies had established contract farming arrangements, and a questionnaire-based survey targeting around 60 contract farmers as well as non-contract farmers was conducted. In addition, in-depth interviews were conducted with selected farmers in each of the villages. Other value chain

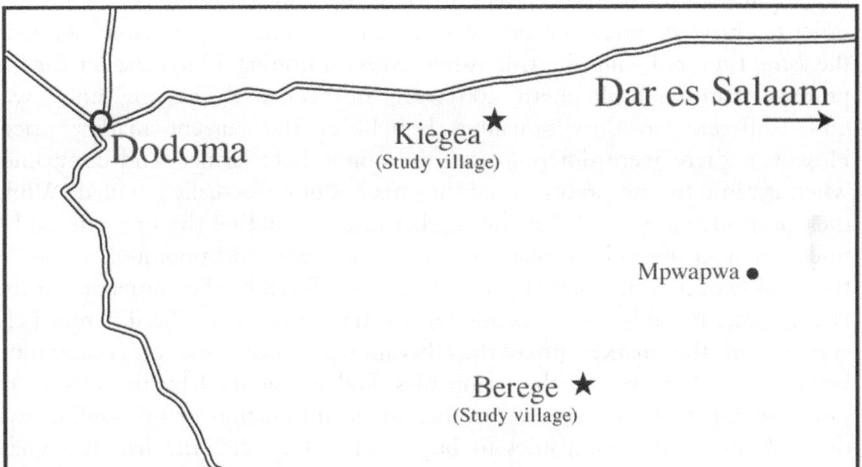

Figure 10.1 Location of the contract farming villages.

Source: National Geographic *et al.* (2011).

actors such as traders and other oil mills were also interviewed to obtain a better understanding of the dynamics of the sunflower value chain.

The contract farming arrangements varied only slightly between the different villages. Farmers were provided with certified sunflower seeds for planting on credit just before the farming season began and were obliged to sell sufficient volumes of seeds to the companies so that they could pay back the agricultural credit. On paper, the contracts stated that farmers were obliged to sell all their production of sunflower seed to the companies, and likewise the companies were obliged to purchase it all from the participating farmers. However, the explanations of how the contract arrangements worked in practice differed substantially from what was put down on paper. One obvious reason for this was the fact that, even though they had signed it, the farmers did not have a copy of the contract in their possession. Apparently no one in the villages had a copy, which resulted in a broad range of perceptions and interpretations among the farmers of what the contracts actually entailed. According to the two companies, the contract arrangements also had to be managed differently from the wording of the contract in order not to accrue large losses, as the farmers would side-sell much of their produce anyway.

Due to the high risk of side-selling, sunflower seeds were supposed to be sold at the market price. Most usually in contract farming arrangements in developing countries, fixed prices are agreed when the contract is signed in order to share the price risk between the farmers and the companies while at the same time reducing the risk associated with finding a buyer/seller for the produce (Eaton and Shepherd, 2001). But in this case the pricing model was quite different, as the companies bought at the current market price. However, there were differences in the opinions of farmers and companies when it came to interpreting what this 'market price' actually entailed. While the companies suggested that the market price should be the one offered by middlemen in the village, many of the farmers were disappointed at this, as they had expected the market price to be that offered by the companies at the factory gate, possibly with a deduction for transport costs. The different perceptions of the market price thus became a major issue of contestation between the farmers and the companies, and it was used by the former to justify selling to third parties. Another main justification of side-selling was the inability of the companies to buy immediately after the harvest, when many farmers are in dire need of cash.

Farmers' trust in the contract arrangements

In a contract relationship, trust is potentially important for both parties. The company needs to be sure that the contracted farmers will keep to the agreement by growing sunflowers according to the instructions and, most importantly, selling their products to the company and paying back their input loans. As enforcement is impractical due to the lack of access to legal institutions, companies need to trust that at least a good proportion of the farmers will honour the agreement so that substantial losses are avoided. On the other hand, the farmers need to trust that the company will buy their products as promised at the agreed price and time. In other words, trust goes both ways without necessarily creating mutuality between farmers and companies. Hence, in principle it is possible to examine how trust affects the operation of the contract farming arrangements from two different perspectives. The first is from the farmers' perspective: why do farmers accept the deal (weighing the risks and rewards, considering social and cultural norms, etc.)? The second is from the perspective of the companies: how do the companies select farmers and villages, taking trust into consideration?

In this study, it was considered important to distinguish between different kinds of trust and how they are produced. The aim is to understand how the trust embedded in the contractual relationship changes over time and what kinds of influence individual behaviour has over economic coordination. For instance, if trust is inherent in rural society or within a community, it might not be so relevant to look at how individuals' behaviour affects trust. Likewise, if intermediaries such as village leaders, schoolteachers or other esteemed and respected individuals in a community are the main catalysts in facilitating cooperation, they would be a more interesting object of study than the contract partners. So the different types of trust that respondents state are important to them in terms of opting for the contract farming arrangements need to be categorised in order to understand the importance of the behaviour of the contracting parties when it comes to building trust.

As briefly mentioned in the first section of this chapter, trust can be categorised into several types, depending on the scale at which trust between two individuals is produced or the types of explanations that individuals give for trusting others. The literature generally distinguishes between 'generalised trust' (often stated to be rooted in social norms and morality) and 'personal trust', that is produced by the specific interactions between individuals (Bachmann, 2006). Note that although both forms of trust play a role in interactions between individuals, the source of the trust differs. Personalised trust can then be divided further into different categories depending on how it is produced:

1 'Process-based trust' implies that the level of trust in the relationship is a result of previous interactions, for example, in an economic relationship where deals have happened over and over again and thereby produced a perception of trustworthiness among the participants. Investigating this type of trust requires questions about actors' relationships with named persons – what have they done in the past in order to earn the trust. A coded example of this type could be this quotation from a farmer: 'You have to check it through time, and wait to see how things turn out, before you can trust him.'

While the above type of trust is produced through interaction, trust can also arise or exist from before any interaction has taken place between individuals.

Not knowing the person you are supposed to trust beforehand requires trust that is not produced by interaction and can come either from an assessment of the character of that particular person or from an assessment of the incentives that act on that particular person. These types will be explained in the following.

2 'Specific character-based trust', follows Moore (1999), although he also includes 'process-based trust' in his category. As one farmer explained it: 'I trusted his speech, he said he would be close to farmers and seemed like a trustworthy person.'
3 'Calculative trust' means that trust is extended to another person based on an assessment of how the trust-taker would benefit from honouring the agreement, that is, would 'calculate' whether it is in his best interests: 'I trusted that the company will come back to do business, because there is not a great deal of competition between buyers in this village.'
4 'Generic character-based trust' relates to trust that the trust-giver is prepared to extend on the basis of characteristics that the trust-taker has because he is a member of a certain group or simply because he is part of society in general (similar to the generalised trust, mentioned above): 'I trust the company because you make an agreement, and you should keep that agreement.'
5 The last category, 'intermediary-based trust', is linked to farmers' trust in a contractor who was 'recommended' by certain important individuals in the village. In this case trust is given by association, and companies may cooperate with trusted locals in order to produce trust between them.

In the empirical interviews, contract and non-contract farmers were asked whether they wanted to continue or begin to participate in the contract farming arrangement or end their participation. They were also asked about the role trust or its absence had in their decision-making. Answers were coded and categorised into the five different categories of trust listed above. In the following, the results of the interviews will be presented to illustrate the importance of farmers having trust in their contract arrangements. Then the results will be discussed in relation to other parameters that might affect participation in contract farming.

The results (Figure 10.2) represent the coding of the notes from each interview. A single interview may contain several coded statements related to trust. 'Positive' or 'negative' represents whether the form of trust production produced a negative or positive outcome. 'Unimportant' represents the fact that a certain form of trust production was not considered to influence whether to trust or not. The results show that the majority of farmers base their perceptions of trust on the experience of engaging with the companies – that is, the dominant form of trust production is process-based. Also, it suggests that farmers do trust the company to a large extent. Respondents explained that initially they were sceptical of the companies'

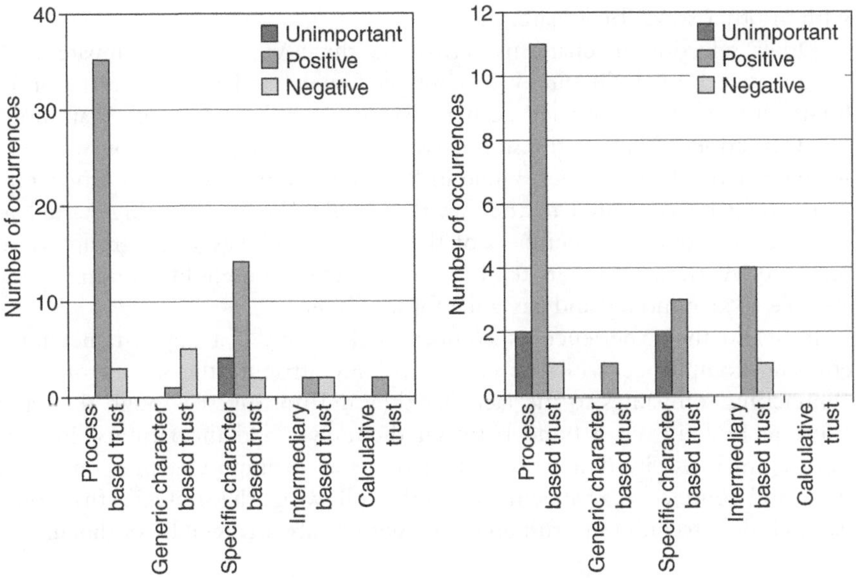

Figure 10.2 Trust categories, derived from interviews with farmers. Left: contract farmers, right: non-contract farmers.

Source: author's fieldwork (2013).

promises, but when the company did what it said it would do, such as providing training and seeds, they came to trust it. There is a generally low level of trust of outsiders, as is evident from the low number of statements supporting the existence of a 'generic character'-based trust. When formal enforcement mechanisms are lacking, it is evident that it is fundamentally important that the company is able to create this trust, either through its actions or in appearance.

It is difficult to assess whether participation in the contract farming arrangement was caused by an inclination to trust one's business partner, as the data were collected after the contract farming arrangement had started. Thus the coincidence of trust levels and participation may be caused either by selection mechanisms (if trusting people are more likely to join) or by the *ex post* build-up of trust. It is clear that other selection mechanisms were also present. For instance the localities of participating households were not evenly spread out through the villages: the participating households had a more central location. Perhaps this location meant that they would hear more about the contract scheme prior to initiation: there were some discriminatory patterns of inclusion due to social relations in the village, especially with the local agent. This important source of non-randomness in contract farming participation might have severe impacts on studies that seek to investigate the income effects of contract farming if the endogeneity problem is not dealt with in other ways (Bellemare, 2012).

Other selection mechanisms across the three villages were apparent. A comparative household survey of contract farmers and non-contract households showed that contract farmers were generally better off than non-contract farmers. The relationship between wealth and participation was only strong in two of the villages, while in Kiegea there was no such relationship. This might be attributed to the fact that, when contract farming started in Chamtumile, only the members of the local SACCO (savings group) were included. As these members tend to have a better household economy, they are able to save money and pay a fee for enrolment.

It is also the experience of companies that trust is a key parameter for ensuring compliance with the contract. They attribute the failure of other contracting companies to the fact that they did not invest enough time and effort in building up a bond between themselves and the farmers. In fact, most companies cite 'being close' to farmers as an important prerequisite of successful contractual arrangements. In the following, the strategies that companies follow to build up trust and cope with its absence will be outlined.

Firms' trust in contract farming

The villages of Berege and Chamtumile were both involved in contract farming arrangements run by the same company for three seasons. Farmers who were invited to participate in the contract farming arrangement in Chamtumile were initially only those who were members of the local Kibakwe SACCO, while in Berege an indiscriminatory approach to participation was adopted. Since SACCO members are usually farmers who are better off economically, participation in contract farming in Chamtumile was more closely associated with wealth than in the other villages. In both Berege and Chamtumile the company had put an agent in place to manage the provision of inputs and the purchase of produced seeds. The choice of agents in both villages fell on those who were already involved in businesses as middlemen or traders. The company owner stated that the reason for choosing these two persons was their previous experience and the perception of them as being 'able to get things done'. From the farmers' perspective, this choice had two implications in terms of building trust. The first was that some farmers were unsure who they were actually dealing with. A few farmers who were asked about the contract farming arrangements were unaware which company was responsible for the contracts and only knew of their agreement with the companies' agent in the village. The second implication was that, in Berege, there was a great deal of mistrust between some of the farmers and the agent. The farmers felt that the agent was distributing inputs in a discriminatory manner and using his position for rent-seeking purposes. Thus using an agent might contribute to the creation of trust on the one hand, due to the existence of a close relationship with the farmers, but on the other hand, if the local agent is not well liked or is perceived to be mismanaging the arrangement, this may undermine the arrangement altogether. Thus creating and maintaining trust through an agent is a strategy the company should be very careful about, since it must ensure that it chooses the right agent in each village.

In the third village, Kiegea, another company had adopted a different approach to managing the contract farming arrangement. The company worked with the local pastor and former village chairman as the local manager. The pastor had no previous experience in trading agricultural products. Even though he was not distrusted as a middleman, as was the case with the agent in Berege, the farmers did complain about his practices. These complaints mainly reflected the perception of discrimination on the part of the pastor, who only invited farmers living in the immediate vicinity of his house, and who were therefore likely to be part of his own social network, to join the scheme; an ox-plough was also distributed in a discriminatory manner. However, this complaint was only voiced by a few farmers. Regardless, the company was quite popular in the village partly because it had distributed seeds free of charge in the first year they operated in the village.

Thus, relations between the farmers and the companies are heavily embedded in pre-existing social relations at the village level, and even though the company may do many things to facilitate the building of trust, social relations in the village continue to play an important role. In general, the creation of trust becomes a strategic venture for some of the companies involved in the contract farming project. Many of the company managers stated that creating trustful relationships with villagers was the key to a successful contractual relationship. Some of the companies use pre-existing relationships with certain villages to ensure compliance. One company based in the market town of Kibaigwa had been trucking food relief to villages on behalf of the government and used this connection as a stepping-stone to involvement with contract farming in specific villages. Another company supplied seeds for planting in the first year as a gesture to farmers. On the other hand, companies still have the power to relocate contract farming arrangements to other villages, an imminent possibility that hovers over the farmers, inducing in them a degree of self-regulation.

Even though farmers are aware that the companies engage in contract farming in their own interests, the language that the companies use to describe their business relations with the farmers reflect what Narayanan (2012) found in India. While company respondents in Narayanan's interviews rephrased contract farming as 'contact farming', the companies in Dodoma talk about 'being close' to the farmers and creating 'bonds' with them. In this moral economy (Clapp, 1994), a certain amount of side-selling is tolerated by some companies when it is caused by acute poverty or crisis in a particular household. While most of the companies indicated that side-selling has declined over the years, when they have engaged with the different villages as trust is being built up, some companies also abandon villages to search for more fruitful economic relationships.

Among the companies that were interviewed, four different types of company strategy were identified in dealing with trust and side-selling. The first was to opt completely out of contract farming due to the perception that farmers were being 'untruthful'. Three companies out of eight decided that their contract farming arrangements earned them more losses than gains and thus returned to sourcing through middlemen and buying at the factory gate as they used to. The second strategy was to scale down operations to concentrate on 'problem-free' villages. In this way the territorial outcome of contract farming is the result of a process of trial and error, the outcome of which is a consequence of the actions of both farmers and companies. The third strategy was to reduce the obligations in the contractual arrangement as a consequence of the company's difficulties with side-selling by farmers and where there was insufficient cash to buy all the farmers' produce. This strategy was taken up by the first company mentioned above, which stopped supplying credit to farmers and also reduced its own obligation to buy all that the farmers produced. In this example, the contract farming arrangement becomes a simpler market relationship, but embedded in the personal relations that the company has built up with the farmers. The last strategy was followed by only a single company, which by contrast expanded the scope of the contract farming arrangement. The company began to supply various other services such as ploughing, spraying and crop insurance, all optional parts of the contract. The company also expanded these provisions to other crops, such as cow-peas, which can be planted alongside sunflower in the same growing season. The company owner explained that the reason for extending the arrangements with farmers was to take advantage of the already existing strong relationships and therefore to avoid difficulties in initiating new activities with them.

Trust as a geographical component

Some GVC authors have tried to employ a more 'geographical' take on the development of contract farming in agricultural value chains in developing countries. Pritchard and Connell (2011) examined the contract farming of chilli in India and found that companies employ different but explicitly geographical strategies when they choose sourcing areas. The companies take advantage of different social structures in the rural landscape that seem beneficial in terms of both the compliance and the productivity of the contracted farmers. In the authors' words, the companies 'deliberately exploited geographical diversity as a stratagem to manage risks under conditions of high uncertainty' (Pritchard and Connell, 2011: 237). The outcome is that rural social structures are either reinforced or challenged as the contract farming arrangements are introduced in the different areas. Hence, the problems of exclusion and participation are thus best understood as interplays between existing territorial features of the rural landscape and the agency of the companies who take advantage of this geography. Also focusing on the problem of uneven territorial development, Fold (2008) shows that international sourcing practices through corporate initiative produce what he calls 'certified spaces' where private regulation rules the market transactions between smallholders and buyers through, for instance, out-grower schemes. Fold (2014) also looks at how settlement trajectories in Ghana and Vietnam depend on participation in specific global agricultural value chains. He argues that the regional development outcomes of different places that are incorporated into specific global value chains can be examined when livelihood studies are integrated with GVC analysis: the territorial embeddedness of the specific global value chain has a profound impact on patterns of uneven territorial development.

As the economic relationships are geographically embedded within the value chain, trust in these relationships also becomes 'geographical'. As shown in this chapter, the inclusion of farmers and villages in contract farming arrangements depends not only on companies' initial choice of sourcing area, but also on the continual production and destruction of trust between farmers and companies. While Fold's (2014) 'privileged spaces' and Pritchard and Connell's (2011) 'geographies of advantages and opportunity' are seen mainly as a condition imposed on rural spaces by the vagaries of relatively more powerful economic actors downstream of them in the value chain, the incorporation of trust and personal relations into the equation results in a more varied picture: the agency of both farmers and companies needs to be acknowledged in processes of (uneven) territorial development. Through contract farming, companies and farmers alike produce or reproduce geographical inequalities in rural Tanzania through an experimental form of economic relation-building. Some villages are able to capture the benefits of contract farming whereby they obtain greater efficiencies in production by using improved technology and thereby earn higher incomes.

Hence, the choice by companies to source from particular villages admittedly entails place-specific inclusion and exclusion, thus creating certain contracting spaces in which farmers have better access to seeds and credit than if they remained outside them. The data suggest that the selected villages have already benefited to some degree from improved infrastructure and various NGO interventions. But while the initial choice of villages is highly uneven, this chapter has shown that repeated processes of building and destroying economic relationships is an important element in territorial differentiation, indicating that territorial inequality might be reproduced through contract farming.

Concluding remarks

Even in a situation where contracts should be able to reduce uncertainty and risk and thus also the need to trust each other, trust continues to play an important role in initiating and sustaining contract farming arrangements in rural Tanzania. This is a relatively banal conclusion, as the institutional framework that should be able to ensure enforcement and compliance is non-existent, but it does imply that policy-makers should be cautious in expecting the benefits of contract farming arrangements that much of the contract farming literature suggests. The formal nature of contracts also assumed in the literature is often not present in the realities of developing countries.

While the value chain literature generally finds personal relations important in more complex forms of economic transaction, where products are complex and non-codifiable (Gereffi *et al.*, 2005), this study has shown that personal relations play a profound role even in the case of 'simple', standardised products such as sunflower seeds. This is due to institutions such as trust being in place and continuously being produced and reproduced in the specific settings, as well as to the lack of formal institutions. Although personal relations are implicitly a part of GVC frameworks, GVC studies do not usually deal with the actual production and distribution of personal relations and trust. Looking at how the production of trust in personal relations between actors can affect the value chain in producing territorial differentiation requires a more specific focus on what trust is and how it is produced and sustained over time, specifically in contract farming relations. Moreover, the processes of uneven development associated with agricultural value chains and the initiation of contract farming arrangements are not only contingent on downstream actors taking advantage of institutions and resources that are unevenly located. They are also a product of the creation and destruction of trust and moral economies between farmers in specific places and actors downstream of them (in this case small and medium-sized companies). Therefore it can be argued that focusing on trust in GVC studies in developing countries may provide valuable insights that can explain the types of coordination that are identified 'on the ground'.

References

Bachmann, R. (2006). Trust and/or Power: Towards a Sociological Theory of Organizational Relationships. In R. Bachmann and A. Zaheer (eds), *Handbook of Trust Research* (pp. 393–408). Cheltenham: Edward Elgar.

Bellemare, M. F. (2012). As You Sow, So Shall You Reap: The Welfare Impacts of Contract Farming. *World Development*, 40(7), 1418–1434.

Clapp, R. A. (1994). The Moral Economy of the Contract. In D. P. Little and J. M. Watts (eds), *Living under Contract: Contract Farming and Agrarian Transformation in Sub-Saharan Africa* (pp. 78–96). Madison: University of Wisconsin Press.

Eaton, C. and Shepherd, A. (2001). *Contract Farming: Partnerships for Growth*. FAO Agricultural Services Bulletin 145. Rome: Food and Agriculture Organization of the United Nations.

Fold, N. (2008). Transnational Sourcing Practices in Ghana's Perennial Crop Sectors. *Journal of Agrarian Change*, 8(1), 94–122.

Fold, N. (2014). Value Chain Dynamics, Settlement Trajectories and Regional Development. *Regional Studies*, 48(5), 778–790.

Fukuyama, F. (1995). *Trust: The Social Virtues and the Creation of Prosperity*. Harmondsworth: Penguin Books.

Gereffi, G., Humphrey, J. and Sturgeon, T. (2005). The Governance of Global Value Chains. *Review of International Political Economy*, 12(1), 78–104.

Granovetter, M. (1985). Economic Action and Social Structure: The Problem of Embeddedness. *American Journal of Sociology*, 91, 481–510.

Harriss, J. (2003). 'Widening the Radius of Trust': Ethnographic Explorations of Trust and Indian Business. *Journal of the Royal Anthropological Institute*, 9(4), 755–773.

Humphrey, J. and Schmitz, H. (1998). Trust and Inter-firm Relations in Developing and Transition Economies. *Journal of Development Studies*, 34(4), 32–61.

Lyon, F. (2000). Trust, Networks and Norms: The Creation of Social Capital in Agricultural Economies in Ghana. *World Development*, 28(4), 663–681.

Moore, M. (1999). Truth, Trust and Market Transactions: What Do We Know? *Journal of Development Studies*, 36(1), 74–88.

Narayanan, S. (2012). *Notional Contracts: The Moral Economy of Contract Farming Arrangements in India*. Mumbai: Indira Gandhi Institute of Development Research.

National Geographic, Esri, DeLorme, NAVTEQ, UNEP-WCMC, USGS, NASA, ESA, METI, NRCAN, GEBCO, NOAA, and iPC (2011). National Geographic World Map, digital topographic basemap of the world. Accessed online via Esri ArcGIS 2015.

Neilson, J. and Pritchard, B. (2009). *Value Chain Struggles: Institutions and Governance in the Plantation Districts of South India*. Chichester: John Wiley & Sons.

Oya, C. (2012). Contract Farming in Sub-Saharan Africa: A Survey of Approaches, Debates and Issues. *Journal of Agrarian Change*, 12(1), 1–33.

Platteau, J.-P. (1994). Behind the Market Stage Where Real Societies Exist Part II: The Role of Moral Norms. *Journal of Development Studies*, 30(4), 753–817.

Pritchard, B. and Connell, J. (2011). Contract Farming and the Remaking of Agrarian Landscapes: Insights from South India's Chilli Belt. *Singapore Journal of Tropical Geography*, 32(2), 236–252.

RLDC (2014). Sunflower Intervention. Online: www.rldp.org/index.php/sectors/8-home/71-sunflower-tanzania. Accessed 20 March 2014.

Warning, M. and Key, N. (2002). The Social Performance and Distributional Consequences of Contract Farming: An Equilibrium Analysis of the Arachide de Bouche Program in Senegal. *World Development*, 30(2), 255–263.

11 Contract farming in a covert sphere

Conspiracy theories as counter-knowledge about sugarcane production in Tanzania

Lotte Isager

Introduction: what is true? What is real?

The theme of this chapter is one particular cultural narrative, which is widely reproduced by sugarcane out-growers who grow and sell sugarcane to two different companies in Tanzania. It is a narrative that describes certain allegedly dishonest and secretive practices on the part of the sugar companies. Hence, it *is known* among the out-growers that the companies manipulate the weighbridge where the tonnage of the sugarcane delivered by individual out-growers is determined. The companies are further *known* to artificially inflate their own costs and thereby manipulate accounts when they calculate the price to be paid to out-growers for their sugarcane. Above all, the companies *are known* to deliberately underestimate the so-called rendement, the main measure of value in the sugarcane, which stands for the recoverable apparent sucrose in it.

According to the out-growers' narrative, the manipulative practices of the sugar companies on the weighbridge, in the laboratory where the rendement is measured and when company accounts are made all result in economic loss for the out-growers and, by implication, economic gain for the sugar companies. Exactly how the out-growers know these things is in many cases difficult to ascertain. Some say they know from personal experience, others say they know from the experiences and stories of others.

This chapter is primarily based on data collected in May and June 2013. As part of an ongoing research project on the contract farming of sugarcane in Tanzania, a total of 40 out-growers – 20 in Location A and 20 in Location B – were interviewed about their experiences with contract farming. During these interviews, the theme of company dishonesty emerged with notable regularity, almost like the chorus of a song. Interestingly, in interviews with leaders of the out-growers' associations and farmers' groups (25 in total), the same theme appeared with great frequency, although a few leaders disagreed with its message. During consecutive rounds of interviews in November 2014 and in May and June 2015, the theme was widely repeated again. It is, in other words, a theme that matters to people.

Representatives and managers from the two sugar companies were also interviewed in 2013, 2014 and 2015. They emphatically rejected the out-growers' claims, not in an aggressive or offended manner, but insistently and with a sense of humour, as if these claims constitute a well-known and utterly correction-resistant discourse on a par with other aspects of their working lives that they cannot control. According to them, the out-growers' allegations have no basis in reality. The companies' official response to the out-growers' narrative is that it is a product of the latter's at times unreasonable and irrational thinking and their tendency to believe in rumours and all sorts of hearsay. At Company A, the management did articulate concern over the mistrust inherent in the out-growers' claims because this was considered unfortunate and not conducive to maintaining an otherwise fruitful collaborative relationship. Apparently, the management was considering how it could convince the out-growers that fraud was not taking place. At Company B, by comparison, although it was an issue the management certainly knew about, it had no intention of changing company practices in order to reduce the level of suspicion among the out-growers.[1]

So what is true and what is real? Are the out-growers suspicious for no reason, as the companies claim? Or do the companies withhold important facts from the out-growers in order to cheat them, as the out-growers say? If something is hushed up in the shadows and omitted from the information that is made available to the out-growers, who then knows these facts – the government? The association leaders? The company managers? Indeed, does anyone know all the facts? These are questions that arise naturally under the circumstances, and they engage a great many people who live the vicinity of the two sugar companies in Tanzania. Out of necessity the questions also engage researchers, since it is clearly of relevance to determine whether the companies are cheating the out-growers or not. The answer will inevitably and fundamentally affect our whole understanding of the realities and future prospects of the contract farming of sugarcane in Tanzania.

As researchers moving in the footsteps of the philosophy of science of the Enlightenment, it would normally be the objective of any study to establish reasonable and coherent links between the facts and to look for causality and explanations – in short, to uncover the truth. However, in this case it would go beyond the practicalities and objectives of the research to try to investigate or inspect the operations of the two sugar companies' weighbridges, let alone their laboratory procedures for measuring rendement or the principles and practices of calculation involved in making up the company accounts. This chapter cannot, therefore, establish the truth of the out-growers' claims against the sugar companies or vice versa. Consequently, the chapter will proceed along a more modest hermeneutic path in search of the meaning(s) of the stories people tell in Locations A and B respectively by treating the out-growers' stories analytically as conspiracy theories. It is crucial to stress at the outset that, in using the term, there is no implication that a conspiracy theory is false by definition, nor is it intended to use the label rhetorically to discredit the out-growers' claims.

Accordingly this chapter employs the pragmatic definition of Sunstein and Vermeule (2009: 205), who define a conspiracy theory as 'an effort to explain some event or practice by reference to the machinations of powerful people, who attempt to conceal their role (at least until their aims are accomplished)'. This definition goes against the long academic tradition of regarding conspiracy theories as discourses of a particularly 'flawed character' (e.g. Popper, 1949; Hofstadter, 1965; Keeley, 1999; Pipes, 1997; for a critique, see Hall, 2006), but it makes possible an analysis of the out-growers' stories in order to learn valuable lessons about the social and cultural conditions of contract sugar farming in Tanzania.

The chapter begins with a brief introduction to conspiracies and conspiracy theories from a theoretical perspective. Following this, the overall conditions of growing sugarcane in Locations A and B are presented. Then, based on notes from interviews with people in the two locations, the content of the out-growers' stories and their allegations against the companies is described in more detail. Finally, conspiracy theories are discussed as meaningful cultural responses to the living and working conditions of the out-growers, which are controlled to a very large extent by incalculable and seemingly inexplicable forces. It is argued that under circumstances where out-growers are denied access to vital knowledge, conspiracy thinking is a method for thinking critically in the absence of a transparent public sphere (cf. Dean, 2000).

Theorising conspiracy theories

The concept of conspiracy theory was coined by Karl Popper in *The Open Society and Its Enemies* (1949). Popper defined it as the view 'that whatever happens in society – including things which people as a rule dislike, such as war, poverty, shortages – are the results of direct design by some powerful individuals or groups'. Popper's main argument was that totalitarian thinking is founded on paranoid imaginary plots in which, typically, monopolists, capitalists, imperialists or 'the Learned Elders of Zion' are given the role of intentionally acting, powerful agents. Popper maintained that conspiracy theories overlook the pervasive unintended consequences of many political and social actions and thereby imply a distorted, pathological form of political reasoning. A similar sentiment was expressed in Richard Hofstadter's dismissal of the 'dangerous and paranoid style' in American politics (Hofstadter, 1965) and, many years later, in Fredric Jameson's notion of conspiracy theories as 'the poor person's cognitive mapping in the postmodern age' and 'a desperate attempt to represent the ... system' (Jameson, 1991: 356, cited in Aupers, 2012: 23).

In the tradition emanating from Popper and Hofstadter, it has long been common in academia to stress the irrational, 'flawed character' (Keeley, 1999) or 'crippled epistemology' (see Sunstein and Vermeule, 2009) of conspiracy theories against the standard of modern, objective and rational science (Aupers, 2012). In contrast to this tradition, Gray takes a more neutral stand and holds that a conspiracy theory should be regarded as a 'counter-discourse of some sort; that is, a discourse seeking to challenge the orthodox or dominant explanation for an event' (Gray, 2010, cited in Sapountzis and Condor, 2013: 2). Dean (2000) similarly stresses the importance of bearing in mind the fact that the social sciences and conspiracy theories actually have many characteristics in common – the attempt to unveil the covert and the deceptive, the urge to reveal what is hidden behind the masks of ideology. Conspiracy theories, 'like other Enlightenment theories … are marked by a drive to know and uncover the truth' (Dean, 2000).

To repeat, the chapter follows Sunstein and Vermeule's (2009: 205) definition of conspiracy theory as 'an effort to explain some event or practice by reference to the machinations of powerful people, who attempt to conceal their role (at least until their aims are accomplished)'. By adopting this definition of conspiracy theories, it is intended to recognise, rather than downplay, the existence of certain similarities between the knowledge-production of the out-growers and the ways in which researchers seek to establish knowledge. It is usual practice in the social sciences to study and reveal how powerful groups in society – sometimes in highly secretive ways – control events and practices, as well as the public representations of these events and practices. As such, the aims of the social scientists may to a large extent resemble the aims of the out-growers. However, since the knowledge-production of the former naturally differs in significant ways from that of the out-growers by strictly following pre-defined criteria and methods of data collection, the analysis of evidence and the drawing of conclusions, the analogy should not be taken too far. It should also be acknowledged, as Hall has rightly pointed out, that 'identifying a conspiracy theory is itself an exercise in power, a reification of power and authority, a legitimation of knowledge, and a tactic to manoeuvre power' (Hall, 2006: 208).

Neither Popper nor subsequent students of conspiracy theories have denied that conspiracies do occur in everyday life. According to Young and Nathanson (2010),

> every real conspiracy has had at least four characteristic features: groups, not isolated individuals; illegal or sinister aims, not ones that would benefit society as a whole; orchestrated acts, not a series of spontaneous and haphazard ones; and secret planning, not public discussion.

Conspiracy theories, for their part, also have distinctive features, above all because of their self-sealing quality; the very arguments that give rise to them and that account for their plausibility make it more difficult for outsiders to rebut or even to question them (Sunstein and Vermeule, 2009: 207; see Sutton and Douglas, 2014, for more on the self-sealing or 'monological nature' of conspiracy theories). Türkay Salim Nefes underlines the political nature of conspiracy theories and suggests that one of the most important characteristics is their attempt to unveil the 'real but hidden' power relations in social groups (Nefes, 2012, 2013). Moreover, conspiracy theories often attribute extraordinary powers to certain agents in terms of their ability to control events, maintain secrets and so forth (Sunstein and Vermeule, 2009: 207).

Several authors agree that the discourse about conspiracy theories has shifted in recent decades from paranoia about an exotic 'Other' standing 'outside' society to paranoia about modern society itself. This modern type is diametrically opposed to traditional forms of conspiracy theories since it is typically about 'the enemy within', about the unknown and malicious forces that operate within the machineries of scientific laboratories, modern corporations, politics and the state (Goldberg, 2000, cited in Aupers, 2012). Likewise, Timothy Melley (2000) observes that 'the term "conspiracy" rarely signifies a small, secret plot anymore ... [Instead,] it has come to signify a broad array of social controls' (cited in Dean 2000: 8). Knight (2000) calls conspiracy theories a transition from 'secure paranoia' to 'insecure paranoia' characterised by inexhaustible suspicion and uncertainty (cited in Aupers, 2012). As the ensuing analysis will show, these theoretical notions are relevant for an understanding of out-growers in Locations A and B. Although the term 'paranoia', with all its negative connotations of psychiatric abnormality, is not suitable for a description of these out-growers in Tanzania, it is still very relevant as a way of emphasising the broader sense of existential insecurity that comes with being a sugarcane-grower on contract.

Growing sugarcane in Tanzania

Company A and Company B are both located in Morogoro region in Tanzania. They were both set up in the early 1960s as nucleus-estate schemes, and the sugar factories continue up to this day to be supplied with sugarcane, partly from estates owned by the companies and partly from out-growers living in the surroundings of the estates. For several years, the companies were managed, not very successfully, as parastatals. In 1998, they were privatised: Company A was bought by a consortium of foreign investors, while Company B was taken over by a group of Tanzanian businessmen, an ownership structure that still persists.

The Tanzanian sugar industry is regulated by the Sugar Board of Tanzania under the legal provisions contained in the Sugar Industry Act 2001. According to the Sugar Act, every out-grower who delivers sugarcane to Company A or Company B is obliged to hold membership of an out-growers' association. At the time of the privatisation, three out-growers' associations existed. However, by 2015, there were a total of 19 associations connected to the two out-grower schemes, two in Location B and 17 in Location A. Contract farming was formally introduced in 2006 in the form of cane supply agreements, which are negotiated every two to three years by the associations and either of the sugar companies. There are no direct contractual bonds between the companies and the out-growers: all mutual communication and financial settlements go through the associations. Hence, the out-growers are not paid directly by the companies but through the associations of which they are members. The companies will transfer payments for delivered sugarcane to the out-growers' association, which will deduct different types of membership fees before transferring the final payment slip to the out-grower.

A sugarcane plant is made up of approximately 80 per cent stalks, from which juice is extracted and crystallised into sugar, and 20 per cent leafy material ('trash') with no value for sugar production. For any sugarcane out-grower, the single most important part of the production cycle is the annual harvest period. The first part of the harvesting process is to burn the sugarcane fields in order to remove the trash. The controlled burning is followed, a day later, by the actual cutting of the stalks, a physically highly demanding job which is done by migrant labourers from other areas in Tanzania. The stalks are then loaded on to trucks and delivered to the factory.

The sugarcane is weighed at the factory weighbridge on arrival, then it is unloaded and given a unique sample number before the procedure for determining the rendement (the recoverable apparent sucrose) in the sugarcane is set in motion. Both companies use the ICUMSA international methods of analysis. A bundle of cane is fed via the cane knives to the first mill with minimal mixing with other consignments. When the cane arrives at the first mill, the 'first expressed juice' is sampled from the first rolls of the mill. This juice is brought to the company laboratory for analysis to determine the rendement, which is expressed as a percentage. The price of sugarcane is always defined per ton with reference to a standard rendement of 10 per cent, and the price received by out-growers for their sugarcane is therefore flexible according to whether the rendement is above or below 10 per cent. Only when this procedure has been successfully completed will the out-grower receive the first payment, which in most cases is vital for the survival and welfare of his or her family in the year to come, that is, before next season's harvest. The amount he gets paid depends on (1) the weight of the delivered sugarcane, (2) its rendement and (3) the price obtained over the following year by the company for the end-product (sugar).

According to the cane supply agreements, the out-growers should receive 90 per cent of the revenue (=harvested volume multiplied by start price) within 30 days of harvesting and the remaining 10 per cent about one year later. The start price is fixed by the companies immediately before the harvest season begins. At the end of the season, when the companies have produced and marketed the sugar, they calculate the final price according to an internationally recognised price mechanism known as 'division of proceeds'. The companies are allowed to deduct several costs related to the production, storage and marketing of sugar when they calculate the proceeds, which are then divided with the out-growers.

What really happens on the weighbridge, in the factory laboratory and in making up the company accounts in Locations A and B is, however, a highly contested story.

The narrative: experiences and explanations of oddities

It is generally known among out-growers that the sugar companies' measurements of rendement cannot be trusted. This understanding is primarily linked to one very common experience, expressed as follows by an out-grower: 'It is strange that sugarcane from the same field, harvested on the same day, can have different rendement going from 12 to 8 or even 7'. An excerpt from another interview further illustrates this experience: Out-grower: 'in 2012, the first truck [from my field] fetched a rendement of 9.0. The next truck from the same field fetched 9.9'. Author: 'Have you asked at the factory why [this could happen]?' Out-grower: 'At the factory they just say that it is normal'.

But no out-grower accepts the official explanation that this could be 'normal'. Instead, they consider it odd, and they have their own explanations for the phenomenon, which can be divided into a number of different categories, as follows.

Stories about the companies' hidden practices

This category of explanation is exemplified in the excerpts from two interviews: Out-grower: [my fields] were harvested in August, but not all was taken to the factory on the first day. So the first day we fetched a rendement of 10.5 per cent. The next day we only got 9.5. It's unfair!' Author: 'Do you think the rendement dropped because one day has passed?'. Out-grower: 'No! It is not because of time, but because the miller takes the average for that day!' Another out-grower said:

> The out-growers do not use fertilisers, so they get a high rendement but a low weight. The estate is the other way round – they use fertilisers and irrigation, and therefore they get a low rendement but a higher weight. The problem is that the investor takes an average of cane from the out-growers and the estate, which means that we get a loss.

The point is, according to the out-growers, that the sugar companies act dishonestly when they mix sugarcane from different sources instead of ensuring, as they were supposed to do, that each out-grower is paid for the amount and quality he or she actually delivers. According to the companies' representatives, it is not normal procedure to mix sugarcane from different sources but under certain circumstances it is unavoidable. One manager from Company B admitted: 'Ideally, we should take samples from each farmer, but that is not possible. We cannot measure for less that 40–50 tonnes, so if farmers deliver less than that, we have to combine with the delivery from another.' Apart from that, he said, the mill and the procedure for measuring rendement function well, 'but you cannot explain that to [the out-growers and the association leaders] if they just sit and listen passively'.

Stories about deliberate cheating by powerful people

Another type of explanation given by out-growers for the same-day/same-field/different rendement problem was delivered by one out-grower with a good-humoured laugh (possibly implying the obviousness of the situation to anyone but the inquisitive researcher):

> I am not an expert, and the chemist in the laboratory is not my relative! If the miller tells him to do anything he can give me a low rendement, from 6 or 7 instead of up of to 12.

Other out-growers claim that it is the associations – not only the companies – that are the culprits: 'The association tries to hide things from us [...] the association might say "this year you will get a rendement of eight percent", but then you get six percent from the miller'. One of the out-growers revealed a payment slip from his association and compared it with the note he had received from the company on the day he delivered his sugarcane. He had delivered two truckloads, one significantly heavier than the other. The bigger load received a higher rendement from the company than the smaller load, but on the association's payment slip the figures were reversed, meaning that the out-grower lost money. This category of explanation clearly includes the understanding that neither the sugar companies nor the association leaders are trustworthy. The particular association in question completely rejected the out-grower's complaint, and for its part the sugar company denied that family relations have any effect on calculating rendement.

Stories about secrecy, uncertainty and the invisibility of the 'real' truth

Yet another type of explanation was given by several interviewees. One person exclaimed:

> If you harvest two acres of cane and load it on two truck trailers, one may get ten percent sucrose but the next only eight percent. How come! We do not know. And we have never been there! We just get the payment slip, and then we can see for ourselves.

Said another out-grower: 'The problem is that the miller does not involve the farmers. The investor is the only one who knows.' And according to a third out-grower: 'We sell our sugarcane, and we just hope for a high rendement. We are not there. Our only option is to believe.'

The out-growers' sense of being disconnected from the truth about how the rendement is measured is equally prominent when it comes to the weighing of their sugarcane and the price calculation that precede their final payment from the companies. It is, in other words, an established truth among out-growers that the sugar companies' weighbridges are not to be trusted. How they know this is in many cases difficult to establish: people just know. However, among those who are willing to explain how they know, several mentioned that one can 'bribe the man at the weighbridge'. None of them admit to having done so themselves, but some say they know of others who use bribes in this way. A few other out-growers claim that '[the company] sometimes calibrates the weight wrongly'. In any case, representatives from both sugar companies categorically rejected the out-growers' claims.

Furthermore, most out-growers express their dissatisfaction with the price determination system in more or less the same words as this out-grower:

> another disadvantage is that the miller is able to report a loss and deduct it from the farmers' payment, but we don't know how this loss occurred, we hear no details, it is all very unclear … If the miller reports extra profit it is not disclosed how the total profit is. We just hear our own amount.

Or, as another out-grower said: '[their] price calculation is unfair … because it doesn't take into consideration the out-growers' cost of production'. Again, the views expressed by out-growers illustrate the experience of being unable to fight and change the companies' intentionally secretive behaviour and the non-transparent ways in which they calculate and account for their expenses. One out-grower expressed the experiences of many people when he mentioned that he had 'never asked about price calculation because [he does] not expect to get a correct answer'.

Truth, secrecy and invisibility

The quotations from out-growers that are cited in the previous section express a feeling that appears to be shared by every out-grower who was interviewed in Location A and, even more so, in Location B, namely a profound sense of being removed from the truth, quite literally. That is, there is a sense of being deliberately denied access to the places and machines where the truth is – the weighbridge, the laboratory and the accounting office – by the companies, which are in full control of these places.

'Secrecy lies at the very core of power', Elias Canetti famously wrote (1981: 290). In interview after interview, people in Locations A and B convey this experience with exclamations like 'How can we know when we are not there?' It is notable that, although such exclamations express the experience of powerlessness against the companies, they still convey a conviction that if only they (the out-growers) were allowed to be present at the weighbridge, in the laboratory and in the accounts office and see the facts for themselves, they would get to know. If only they were allowed to be there, things would be transparent, and they would be able to understand the procedures for the weighing, the measurement of rendement and the price calculation – read the weight, extract the juice, analyse it chemically, interpret the results, do the price calculation. This scenario, in other words, contains an image of the companies as very powerful 'Others' – and enemies – who stand outside and against the community of the out-growers. Within this scenario, the out-growers' problem is of a purely epistemological nature: they do not know what is true.

In contrast to this scenario, however, it is important to note that there are also out-growers whose sense of powerlessness vis-à-vis the company reflects a different and arguably a deeper experience than 'merely' being refused access to certain locations and facts. To exemplify this, one out-grower stated that, '[even] if I went to the factory and saw how they measure the rendement, I might still not understand it'. Another out-grower explained at length that in the past the companies used 'a better method':

> When the truck was weighed, it got two code numbers for the first and second trailer. When those numbers were put into the computer the pulling machine would begin to pull the cane in, and when the first ones were crushed, an alarm bell would ring and a small amount of juice would be extracted and sent to the laboratory ... Even though the farmers were not there, the process was open ... we could understand ... it was done manually ... [We] could see people bringing juice and code numbers ... [but nowadays,] you cannot see what is going on.

Later in the interview, it turned out that this out-grower had never actually seen the old process, but he 'could imagine it', he said, unlike the current system of measurement. And when asked when the old system was replaced by the current system, he replied: 'It happened in secret. They have never told us anything about it.'

What this latter group of out-growers express is a sense of personal and communal insecurity in the face of modern production mechanisms, which are perceived to be utterly beyond not only their control but also their comprehension. For these people, the 'enemy' that emerges in their discourse is not just the sugar companies in the forms of secretive, sometimes plotting, always powerful 'Others' standing outside the out-growers' own community, with absolute control over every important aspect of the company–farmer relationship. Instead, the enemy 'Other' is perceived to be within themselves as well in the form of their sense of insufficient brain power or education. To the extent that this sense of powerlessness is associated with a degree of paranoia, it is the 'insecure paranoia' characterised by inexhaustible uncertainty, rather than the 'secure paranoia' associated with more traditional forms of conspiracy theories (cf. Knight, 2000, cited in Aupers, 2012). In other words, this is a scenario where out-growers face not only epistemological but also what might be called ontological insecurity: they do not know what is true, and on top of that they also come to doubt what is real.

Contract farming in a covert sphere

The company managers who were interviewed, by contrast, were resolute in refuting the allegations of dishonest practices on their part. They are aware of the out-growers' complaints and express their regrets over the out-growers' belief in hearsay, as opposed to the ostensibly more rational explanations presented by the companies. As already mentioned, it is beyond the scope of this chapter to uncover the truth of these matters. However, it is reasonable to treat the out-growers' conspiracy theories as culturally meaningful 'strategies to rationalise anxieties by developing explicable accounts for seemingly inexplicable forces' (cf. Aupers, 2012: 28). When the out-growers insist on making sense of repeated events and practices that act to their disadvantage by referring to the machinations of powerful company managers and other influential people who attempt to conceal their role, it is not because of any 'crippled epistemology' or irrational 'flawed character' but because, under the circumstances, it is the most sensible and meaningful explanation of these events and practices.

Both companies maintain complete control over weighing procedures, the process of measuring rendement and determining the start price and final price paid to the out-growers. Inherently, the companies fully control what information is to be shared with the out-growers in connection with these vital production and business processes, this being stated in the cane supply agreements. Apart from defining the rights and obligations of the companies and the associations, these 20-odd-page contracts contain details about harvest procedures, payment procedures, lists of deductible company costs, etc. They are written in English in legalistic jargon and are generally considered difficult to comprehend by association leaders.[2] Out of 19 association leaders who were asked what influence they have over the content and design of the cane supply agreements, only one replied that he had such influence. Seven others felt they had 'some influence', but did not specify what exactly made them say so. The remaining 11 leaders said 'no influence at all'. None of the out-growers who were interviewed had ever seen the cane supply agreements, and they did not know their content. Several were unaware that such contracts even existed.

It is interesting to contemplate how not only out-growers but also association leaders experience the limitation of their powers. Thus, one association leader explained in an interview that 'We just receive information about the price and cannot argue … the miller uses the world market as an argument'. Another association leader dug into a drawer in his desk and found the official lists of company expenses deducted by the two companies before the calculation of the final price paid to out-growers in the 2011–12 season. Comparing the two lists, one for each company, the association leader noted that Company B spent 90,000 Tsh/ton on transport to depot, whereas the equivalent for Company A was 35,000 Tsh/ton. Company B also deducted 70,345 Tsh/ton for so-called discounts and commission, whereas Company A had no such items in its list of company expenses. 'So what can we do?' asked the association leader; 'the [Company B] owner uses his own transport company to carry our sugar ... he can charge whatever he wants – it is his own company!'

The unequal power relations between companies, out-growers and associations are indisputable. It is likewise incontestable that, for the out-growers, several things that are largely beyond their control can go wrong in connection with the harvest process. First, not every out-grower manages to harvest every season because the companies' crushing capacities are insufficient relative to the delivery capacities of the out-growers and because of favouritism among the association leaders who are in charge of managing the harvest for their members. Many out-growers report that they need to bribe their association leaders to ensure that their fields are selected for harvesting. Frustration over these conditions has caused many out-growers to form new associations in recent years in Location A. Second, bad infrastructure, weather conditions and broken machines may all prevent sugarcane being transported to the factories. Third, uncontrolled fires can cause disaster. If sugarcane is left uncrushed after burning and cutting, the rendement will naturally diminish after two or three days, and the cane becomes almost worthless. Therefore, smooth harvest operations by the associations and the contractors that do the actual cutting and transporting of the sugarcane is critical, but smooth operation is often not achieved, resulting in economic losses for the out-growers. Fourth, even if harvest procedures are completed faultlessly, the out-grower may not receive his or her payment within the time period stipulated in the cane supply agreements. Hence, in Location B, payment delays of up to five months are commonplace. In Location A, delays are less frequent and much shorter, but they happen nonetheless (see Magongo, 2008; Ngirwa, 2011; Smalley *et al.*, 2014). Complaints over payment delays have been filed with the Sugar Board of Tanzania by out-growers and their associations numerous times, but due to regulations contained in the Sugar Act 2001, the Board is unable to intervene in this crucial matter. At the same time, the Sugar Act prohibits out-growers from selling their sugarcane to other buyers than the two companies.

Officially, contract farming of sugarcane is a success story in Tanzania. Since Company A and Company B were privatised in 1998, the total volume of cane crushed annually has doubled at Company B and tripled at Company A. Between 1998 and 2012, the number of out-growers and the acreage of out-growers' fields at Company B and Company A have also doubled and tripled respectively (company data, obtained May and June 2013). Prices have gradually increased since 2000, particularly in Location A. Further expansion of Tanzania's sugarcane production is planned in connection with the 'southern agricultural growth corridor' (SAGCOT), just as the expansion of contract farming more generally is included in the country's agricultural development vision *Kilimo Kwanza*'s promise to deliver institutional capacity-building and (fast-track) legislation to support contractual systems. But seen from the perspectives of out-growers in Locations A and B, this apparent national success story conceals local controversies and disasters, which tend to be omitted from the official discourse. After years of complaints about prices and payments, out-growers in Location B are uprooting their sugarcane in large numbers, and their share of sugarcane delivery has dropped dramatically. Company B is at the same time expanding its estate lands considerably, possibly aiming to eliminate its dependence on out-growers. In Location A out-growers are not facing economic disaster on the same scale as in Location B, but the limited crushing capacity of Company A in comparison with the delivery capacity of the out-growers still means that, every season, hundreds of out-growers fail to harvest, which has obvious catastrophic consequences for them.

Conclusion

Under the circumstances outlined above, the out-growers' conspiracy theories function as a kind of counter-knowledge against powerful representations of the sugarcane production in the two areas. Out-growers are prevented by the sugar companies and the associations from accessing knowledge that is vitally important to them. Following years of dealing with secretive and, at times, corrupt associations and a government that is unable or unwilling to help them when, for example, the sugar companies break their contracts by delaying payments, the out-growers are forced to reflect upon and make sense of their lives in the absence of a transparent public sphere.

It is clear that the out-growers' stories express a longing for epistemological certainty, as well as a sense of having a right to it in the form of access to accurate knowledge about the weight and rendement of their sugarcane, etc. They want to know what is true. It is perhaps more thought-provoking that, by revealing the forged reality of official discourses about the contract farming of sugarcane in Locations A and B, the out-growers' conspiracy theories also express a longing for ontological certainty and for a social consensus over 'what is real'. As such the out-growers' conspiracy theories demonstrate a forceful and compelling will *to know* in the way an ideal public sphere would allow people to know. But as contract farmers in a covert sphere, the out-growers' desire for transparent knowledge is severely and deliberately denied them by forces too strong to combat. In this situation, it would have been nice if researchers could assist the out-growers and disclose the hidden facts. Naturally, researchers share the out-growers' desire for knowledge and to uncover the truth, as it is a fundamental task of science and research to put knowledge into the public domain. But as already noted, in this case research is not able to live up to its own ideals. And although out-growers in Tanzania receive researchers with great courtesy and friendliness, it is notable that researchers seem to fall into the category of not-necessarily-to-be-trusted people who have come into the out-growers' lives in recent years. One out-grower summed up the experience shared by many other out-growers when he stated that researchers 'come to the village and ask the same questions … then they leave again, [and] we hear nothing'. Unfortunately, this is very likely to be true, as any researchers do indeed come to extract data for their own work and leave again without further ado. The out-growers' conspiracy theories should therefore be understood as a method for thinking critically (cf. Dean, 2000) in a social and cultural context dominated by powerful 'Others' who control information flows in their own and not the out-growers' interests.

Notes

1 It is beyond the scope of this chapter to discuss the two companies' different approaches to this topic.
2 In 2015, a Swahili translation was made available for the first time.

References

Aupers, S. (2012). 'Trust No One': Modernization, Paranoia and Conspiracy Culture. *European Journal of Communication*, 27(1), 22–34.

Canetti, E. (1981). *Crowds and Power*. New York: Farrar, Straus and Giroux.

Dean, J. (2000). Theorizing Conspiracy Theory. *Theory & Event*, 4(3). https://muse.jhu.edu/article/32599.

Hall, J. A. (2006). Aligning Darkness with Conspiracy Theory: The Discursive Effects of African American Interest in Gary Webb's 'Dark Alliance'. *Howard Journal of Communications*, 17(3), 205–222.

Hofstadter, R. (1965). The Paranoid Style in American Politics. In R. Hofstadter (ed.), *The Paranoid Style in American Politics and Other Essays*. New York: Knopf.

Keeley, B. L. (1999). Of Conspiracy Theories. *Journal of Philosophy*, 96(3), 109–126.

Magongo, J. K. (2008). Research Paper in Partial Fulfillment of the Requirements for Obtaining the Degree of Masters of Arts in Development Studies. The Hague, Institute of Social Studies.

Melley, T. (2000). *Empire of Conspiracy: The Culture of Paranoia in Postwar America*. Ithaca, NY and London: Cornell University Press.

Nefes, T. S. (2012). The History of the Social Constructions of Dönmes (Converts). *Journal of Historical Sociology*, 25(3), 413–439.

Nefes, T. S. (2013). Political Parties' Perceptions and Uses of Anti-Semitic Conspiracy Theories in Turkey. *The Sociological Review*, 61(2), 247–264.

Ngirwa, A. K. (2011). Contractual Governance in Agro-Industry Institutions in Tanzania. Unpublished PhD Thesis. Rijksuniversiteit Groningen, Netherlands.

Pipes, D. (1997). *Conspiracy: How the Paranoid Style Flourishes and Where It Comes From*. New York: The Free Press.

Popper, K. (1949). *The Open Society and Its Enemies*. Princeton, NJ: Princeton University Press.

Sapountzis, A. and Condor, S. (2013). Conspiracy Accounts as Intergroup Theories: Challenging Dominant Understandings of Social Power and Political Legitimacy. *Political Psychology*, 34, 731–752.

Smalley, R., Sulle, E. and Lameck, M. (2014). *The Role of the State and Foreign Capital in Agricultural Commercialisation: The Case of Sugarcane Out-growers in Kilombero District, Tanzania*, Working Paper 106, future-agricultures, Institute for Poverty, Land and Agrarian Studies (PLAAS), University of Western Cape, South Africa.

Sunstein, C. R. and Vermeule, A. (2009). Symposium on Conspiracy Theories. Conspiracy Theories: Causes and Cures. *The Journal of Political Philosophy*, 17(2), 202–227.

Sutton, R. M. and Douglas, K. (2014). Examining the Monological Nature of Conspiracy Theories. In J. W. van Prooijen and P. A. M. van Lange (eds), *Power, Politics, and Paranoia: Why People are Suspicious of their Leaders*. Cambridge: Cambridge University Press.

Young, K. K. and Nathanson, P. (2010). *Sanctifying Misandry: Goddess Ideology and the Fall of Man*. McGill-Queen's University Press.

12 Does contract farming empower smallholder agricultural producers?

Lessons from sunflower contract farming in Tanzania

Joseph Andrew Kuzilwa and Daniel Mpeta

Introduction

This chapter revisits the concept of smallholder empowerment and argues that economic outcomes of contract farming are enhanced if they are accompanied by what is referred to as political empowerment of the contract. Research on contract farming involving smallholders in sub-Saharan Africa has mostly focused on the economic drivers and outcomes of such contracts. Using methodologies that adjust for possible selection bias, a number of these studies have shown that contract farming enhances productivity and efficiency, thereby increasing incomes and overall welfare (see, for example, Bellemare, 2012; Cai *et al.*, 2014; Key and Runsten, 1999; Mpeta, 2015). These outcomes result from contract farming interventions in areas that are considered problematic for smallholders, like access to improved inputs, extension services and markets. The outcomes certainly empower the smallholders economically.

However, scepticism has been expressed regarding the contracts' institutional arrangements and the extent to which they lead to an optimal realisation of economic outcomes. One of the arguments is that the economic outcomes of contract farming arrangements can be enhanced even further by addressing the asymmetrical power relations that characterise many contract farming arrangements in Africa, where the interests of smallholders, who are perceived as the weaker party in the relationship, are subordinated in the contractual arrangement (Utting, 2012). These asymmetrical power relations influence such aspects as access to information, negotiating power with other market actors and participation in the design of contract farming arrangements. In order to cushion smallholders involved in contract farming schemes from these impediments, so-called political empowerment has to be initiated.

This chapter develops an analytical framework to examine the economic and political empowerment of contract farming and adopts a framework with which to assess the efficacy of sunflower contract farming in Kongwa District in the central parts of Tanzania. Contract and non-contract participating smallholders are compared to assess whether the absence of political empowerment affects the realisation of economic outcomes. The chapter begins with an elaboration of the notion of empowerment and contract farming as a means to empower smallholders, followed by a presentation of the analytical framework, which draws on Utting's (2012) empowerment model. Then the economic and political empowerment of sunflower contract farmers is analysed according to a number of variables, followed by a discussion of the resulting economic outcome. The chapter ends with some concluding remarks.

Revisiting empowerment in contract farming

As a concept, empowerment is subject to different definitions depending on different academic disciplines and geographical contexts. The application of the concept has ranged from organisational management studies (Bandura, 1997; Conger and Kanungo, 1988) to rural development (Narayan, 2005). The notion of empowerment has often had political connotations, especially when it is applied to the political and economic independence of emerging economies. For example, it has been related to self-direction, liberation, autonomy, self-determination, the capacity to fight for one's rights, independence, one's own decision-making, being free, awakening, self-strength, capability participation, control, own choice, self-confidence and mobilisation (Ibrahim and Allen, 2012; World Bank, 2002; Narayan, 2005; Mushumbusi and Kratzer, 2013).

During the late 1990s, the World Bank advocated the concept of empowerment and the notion of farmer empowerment is now an integral part of many international and national development agencies' policies for supporting agricultural and rural development (World Bank, 2002). A central argument for supporting farmer empowerment on the part of development agencies is that there is a strong relationship between farmer empowerment and development outcomes such as poverty reduction and agricultural growth. In addition, farmer empowerment is considered important for demand-driven and market-oriented agricultural service provision, as it may increase market opportunities, but also potentially expose and catapult smallholders into relationships where they have to negotiate with more powerful actors in order to obtain commercial terms that will improve their economic position. As individuals, smallholders are the weaker party in negotiations with, for example, private buying companies or their agents.

Farmers' groups are one of the key mechanisms for supporting farmer empowerment. It is generally accepted that a farming community must be organised in some sense if smallholders are to improve their ability to demand and access services from public and private agencies and to negotiate with greater strength in agricultural input and output markets. To some extent it is assumed that farmers' groups are capable of 'standing alone' in the market and that they have sufficient market presence (volume) and capacity to strengthen their bargaining power.

However, the definition of empowerment and what farmer empowerment means vary significantly among development agencies and academics. Petesch *et al.* (2005: 40) define empowerment as 'increasing both the capacity of individuals or groups to make purposeful choices and their capacity to transform these choices into desired actions and outcomes'. Empowerment is said to have happened 'when people, individually or collectively, conceive of, define and pursue better lives for themselves' (quoted in Prato, 2013: 2). Hence, as Petesch *et al.* (2005) emphasise, empowerment can be defined at both the individual and group levels. The World Bank (2002) defines empowerment more broadly as 'the expansion of assets and capabilities of poor people to participate in, negotiate with, influence, control, and hold accountable institutions that affect their lives' (World Bank, 2002: xviii). While recognising that there is no agreed understanding of empowerment or unified empowerment strategies, the World Bank (2002) has identified four intertwined elements that are considered important for assessing the presence and extent of empowerment: access to information; inclusion and participation; accountability; and local organisational capacity.

Access to information is described as 'information that is relevant, timely, and presented in forms that can be understood' (World Bank, 2002: 15). In relation to contract farming, this could refer to the information in the contract, but also information on extension services and markets. Inclusion and participation refer to the kinds of development interventions in which people are included, but also how they are included. For example, how are smallholders selected in contract farming arrangements and under what terms? Accountability is the ability to hold public officials, private employers or service providers to account for their actions or mandates (World Bank, 2002: 16). In relation to contract farming arrangements, this could be a question of whether contracting firms can be held accountable by smallholders for complying with contract terms. The assumption here is that smallholder farmers are organised and have the powers to do this. Hence, local organisational capacity refers 'to the ability of people to work together, organise themselves, and mobilise resources to solve problems of common interest' (World Bank, 2002: 17), whether with or without formal rules and organisations. In contract farming arrangements, this could refer to the existence of farmers' associations and the capacity they have to negotiate and execute contract terms.

This definition of empowerment complements work by Utting (2012), who looks at empowerment from two different perspectives: the economic and the political. According to Utting, economic empowerment entails access to drivers of production and the resulting economic outcomes, while political empowerment entails capacity and actual involvement relating to contractual negotiations and dispute resolution. Key and Runsten (1999) also argue that the type and amount of benefits acquired by smallholders depend on the strength of their bargaining power. Such benefits include access to adequate input supplies needed in production, including supplies of the required volume of improved seeds, access to adequate credit, direct production assistance (for instance, in terms of access to tractors) and a 'fair' product price. However, when left on their own, smallholders often have limited bargaining power when they deal with more powerful and better informed agents such as processors and/or buyers of farm produce. Global value chains often involve important power asymmetries between the actors involved. Indeed, while entry into global value chains may offer new opportunities for improvements to farmers' livelihood, without specific external support to their empowerment smallholders are often left on their own, despite being organised in farmers' groups (FAO, 2011).

While recognising that contract farming is not a one-size-fits-all solution to the challenges of smallholder production and marketing, certain features are considered critical. De Schutter (2012) observes that, in assessing any contract arrangement, one should not just look at whether it is economically viable for both the buyer and the farmers, but also at whether negotiations are on an 'equal footing', since this is certainly influenced by bargaining strength and information symmetry. Drawing on the definition of empowerment above, contract farming may be considered to be a means to empower smallholders if farmers themselves are involved in contract negotiations and price-setting while also gaining access to credit and improved agro-inputs, extension services and assured markets. Taken together, this could result in the enhancement of productivity, market access and increased competitiveness.

In the case of many cash crops grown in sub-Saharan Africa, contract farming serves to integrate farmers into higher-value markets within the global value chain. As well noted by the global value chain literature (e.g. Gereffi, 1999; Gibbon, 2001; Fold and Larsen, 2011), the magnitude of the gains of participants in global agro-industrial value chains depends on the chain's governance structure and the roles played by the actors within it. Some strongly export-oriented value chains like tobacco are dominated by large transnational corporations and tend to concentrate value at the final processing and consuming end of the chain (Utting, 2012). Essentially the same kind of domination, but of a different magnitude, may be exercised by a local firm, as with the sunflower chain in Tanzania (see below). In other words, the value additions and benefits that accrue to the smallholders may be compromised if the farmers are not appropriately represented, their voices are not heard and an implementable dispute settlement procedure or mechanism is not established. In this context, it is apparent that economic and political empowerment are complementary aspects of one system and need to be assessed together when examining the implementation and economic outcomes of a particular contract farming scheme involving smallholders.

The following analysis of farmer empowerment in sunflower contract farming arrangements in Tanzania takes as its point of departure institutional economics and Utting's (2012) empowerment framework. As illustrated in Figure 12.1, smallholders' economic empowerment includes a number of aspects that contribute to feasible economic outcomes, namely access to inputs (improved seeds), credit, market, technology, information and extension services.

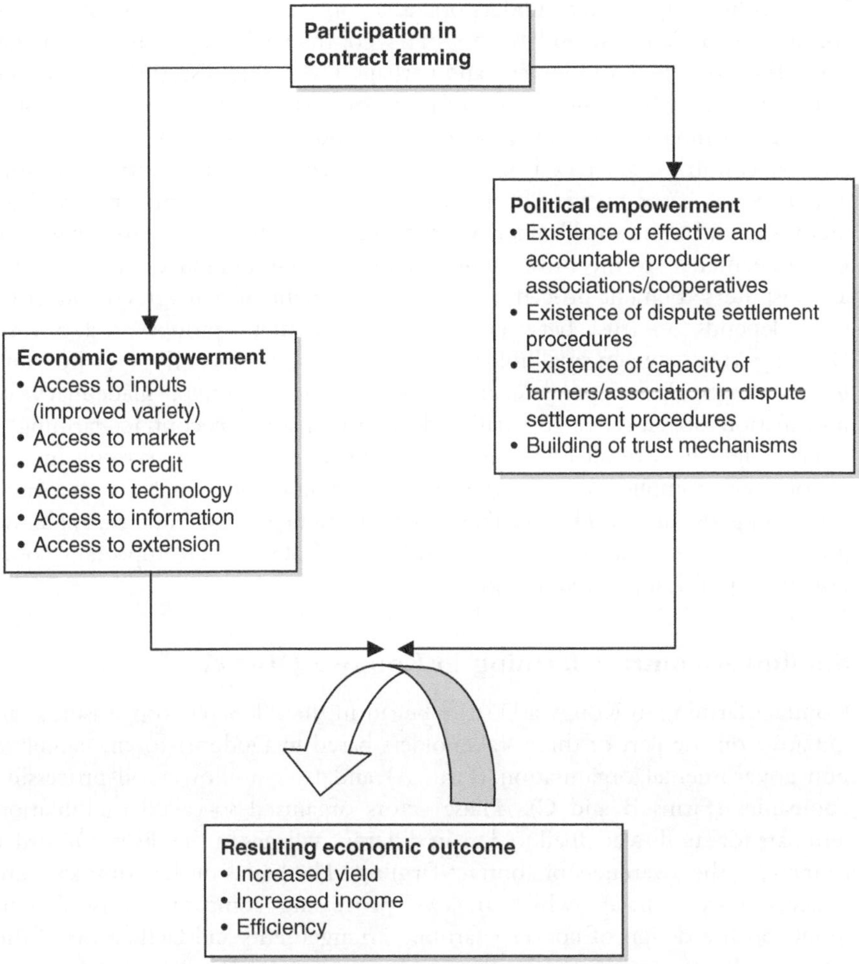

Figure 12.1 Participation in contract farming and farmers' empowerment.

Source: own construction based on Utting (2012).

The notion of the political empowerment of smallholders embraces (1) the existence of effective and accountable farmer groups, (2) the existence of dispute settlement procedures and (3) the ability of individual smallholders and/or farmers' groups to access legal advice in dispute settlement procedures – for example, in negotiating and designing the contents of the contract (Utting, 2012).

In addition to the analytical dimensions of political empowerment provided by Utting (2012), we include the building of trust mechanisms into the analytical framework as another aspect of political empowerment. We argue that Utting's (2012) empowerment model assumes that contract farming is based on 'formal contracts' understood as a contract signed by farmers' groups (or individual farmers) and buyers. The contract stipulates the contractual obligations of sellers and buyers, and perhaps also includes sections on dispute settlement procedures in the case of non compliance. However, while some contract farming systems in Tanzania, and indeed some other sub-Saharan African countries, are based on detailed written contracts, contract farming arrangements based on very simple elements (e.g. a list of inputs received on credit signed by the smallholder or other seller and the buyer respectively) or oral agreements are the most common. In these circumstances, and without any dispute settlement procedures, the enforcement of oral agreements critically depends on trust between farmers (or farmers' groups) and buyers. Building trust between smallholders and buyers creates incentives for honouring contracts (Doney and Cannon, 1997). For example, inadequacy of information of relevance to smallholders participating in contract farming – which might be a result of asymmetrical power relations – may result in suspicion and complications in the business transactions, such as increased side-selling by smallholders (Zakaria, 1999). Hence, trust built through the provision of, for example, information to smallholders is an important component of political empowerment.

Sunflower contract farming in Kongwa District

Contract farming in Kongwa District began in the 2008/09 crop season as an initiative on the part of three stakeholders based in Dodoma town, namely a non-governmental organisation (Firm A) and two sunflower oil-processing companies (Firms B and C). These actors organised so-called mobilisation seminars for sunflower smallholders in different villages in the district in order to increase the awareness of contract farming. The basics of the contract were conceived by Firm A, while the two processing companies assisted with inputs on the design of contract farming arrangements and facilitation of the seminars. It was anticipated by the project's initiators that the education and sensitisation would reduce risks and increase the viability of contractual arrangements.

Through the seminars, the companies proposed that smallholders cultivate sunflower in the area and participate in the contract farming scheme. The information and knowledge provided at these seminars convinced some farmers to do so. Most farmers were motivated by the conditions entailed in the contract in which they would be guaranteed a market outlet for their sunflower seed production and support in the form of inputs and farm implements such as seeds, ploughs and canvases. The attractiveness of sunflower production by smallholders was also enhanced by the government's recent focus on promoting sunflower as one of the major oilseed crops in Tanzania, in contrast to the earlier situation, when it was largely considered a traditional and unimportant crop. During the mobilisation seminars, and as part of the contract terms, the sunflower oil-processing companies promised to establish a sunflower marketing centre in the villages, facilitate the formation of farmers' groups, buy all the sunflower seed the farmers produced at the going spot-market price, facilitate and finance farmers' demonstration plots in the selected villages (e.g. Ihanda), extend appropriate knowledge through seminars on proper sunflower agronomics and contract farming arrangements, provide a credit of 5 per cent of what the farmer would need for sunflower husbandry activities (e.g. land preparation and purchase of fertilisers) and provide improved seed variety at cost. The contractual system and contractual conditions are shown in Figure 12.2.

In turn, farmers were required to produce more sunflowers and sell at least half their total production to the contracting firms. They would also be required to comply with all other agreements, such as planting on time and observing certain agronomic practices. Two important things are worth noting. First, the contractors invited all the farmers to join. There were no selection criteria and participation was open to all interested farmers – the

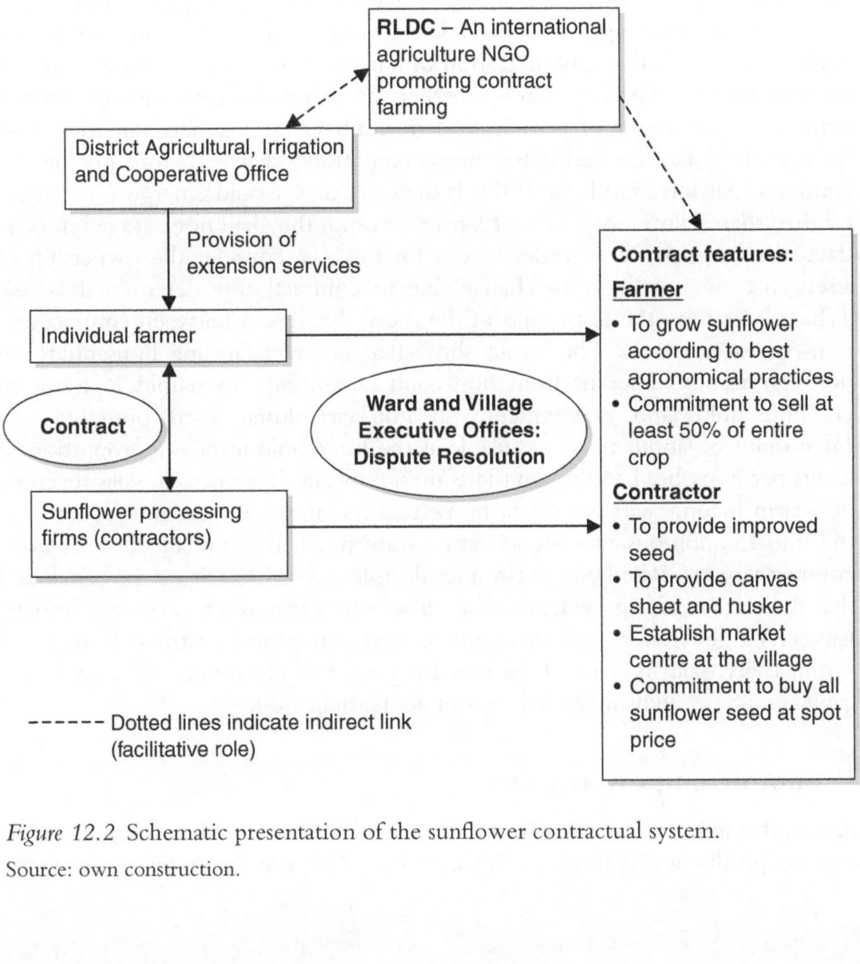

Figure 12.2 Schematic presentation of the sunflower contractual system.

Source: own construction.

terms of the contract determined how the farmers were expected to participate. Second, unlike many contract arrangements, the sunflower contract was not based on any guaranteed floor price or quality premium. Among most of the smallholders, contract farming was considered to be an arrangement that represented a practical framework for access to production and marketing facilities, and many farmers adopted the idea with great interest when it was introduced.

The following two sections assess smallholder contract farmers' economic empowerment and political empowerment according to the framework depicted in Figure 12.1. The analysis of these forms of empowerment and the resulting economic outcomes are based, first, on in-depth qualitative interviews with smallholder sunflower farmers and the sunflower oil-processing companies and, second, on a quantitative survey of 400 contract and non-contract sunflower smallholders in Kongwa District in 2012.

Economic empowerment is analysed by comparing the economic empowerment variables between contract and non-contract farmers and testing their significance. Statistically significant differences in what are considered aspects of economic empowerment would suggest that contract farming does make a difference in empowering smallholders. Using simple t-test results, an attempt was made to assess whether any indication of self-selection bias is observable among contract farmers. The question is whether the observable gains among contract farmers are the result of participation in contract arrangements or they exist because these farmers had better initial conditions, that is, before joining the contracts (self-selection bias). If this is the case, they would emerge as far better off than their fellow non-contract farmers, even if they had not engaged in contractual arrangements. In order to test for this, we consider the ownership of assets that are not likely to change due to contracts (pre-determined assets). Table 12.1 shows the test results of the mean differences between contract and non-contract farmers. The results show that contract farming households are not significantly different from non-contract farming household in terms of (1) household land endowment, (2) non-agricultural assets possessed and (3) amount of labour power available at the household level as a proportion of adults per household. Although there are significant differences in sales revenue, non-farm income and per capita household income, this does not suggest that the initial conditions of contract farmers are much better than those of non-contract farmers. Based on the t-test results (see Table 12.1), it can be concluded that there is no strong evidence of a self-selection bias among contract farmers. Observable differences between contract and non-contract farmers in terms of productivity, sales revenue, total non-farm income and household income per capita could be safely attributed to contract participation.

Economic empowerment

As mentioned above, economic empowerment is defined here as access to various production drivers, in this case including inputs, credit, technology, markets, information and extension services. The descriptive statistics and simple selection bias test (see Table 12.1) compare contract and non-contract farmers and indicate that participation in the contract farming of sunflowers results in better economic outcomes than non-contract participation. Elements of economic empowerment are therefore detected when the two groups are compared.

Access to improved seeds

Although sunflower cultivation for oilseed production requires improved seeds, fertilisers and other inputs for quality production, only improved seed is provided to farmers through the present contractual arrangements. On average, those farmers who obtain inputs on contract have better access to higher-yielding, quality declared seed, and they have a higher use of improved seed (46 per cent) compared to non-contract farmers (9 per cent). Still, contract farmers reported having received less certified seed than they actually needed. In certain areas, a fixed amount of seed (only 3 kg) was distributed to contract farmers regardless of farm size. Consequently, farmers are forced to use lower-yielding local seed varieties to supplement their seed needs (see Table 12.2). About 54 per cent of the contracted farmers could not secure a sufficient quantity of quality seed and had no other option but to top up with local seed varieties to satisfy their total needs. Local varieties are either obtained cheaply from local markets or come from the smallholder's previous harvest stock.

About 80 per cent of the contract farmers mentioned the need for better-yielding seed as a major economic motivation for adopting contract farming, market access (15 per cent) and extension services (5 per cent) being of far less importance. Better access to quality seed can thus be considered to be the most important aspect of economic empowerment.

Access to credit

During focus-group discussions, farmers unanimously raised the issue of poor access to credit to pay for services like, for instance, tractors and ox ploughs for cultivation tasks and the procurement of fertiliser. None of the contractors (sunflower oil-processing companies) had provided cash credits to facilitate sunflower production, and only 4 per cent of all farmers reported access to credit services outside the contractual arrangements. Those who had access to credit services named informal savings and credit schemes run by villagers themselves on a self-help basis as the sources of credit. These loans were relatively limited and generally very small, no more than a few thousand shillings. The lack of credit facilities also hinders farmers from hiring other critical production inputs like labour, which is especially needed for seasonal labour-intensive activities like weeding and harvesting.

Table 12.1 Summary statistics of socio-economic characteristics of sunflower farmers

Variables		*Total sample (means)*	*Non-contract farmers (means)*	*Contract farmers (means)*	*T-test of means difference*
Household	Age	42.55	41.36	43.67	–1.81**
	H/h size	5.38	5.44	5.33	0.508
	Adults-ratio	0.496	0.486	0.506	–0.906
Farm	Farm size	3.6	3.4	3.7	–0.2933
	Farm-ratio	0.48	0.47	0.49	–1.0680
	Land-endow.	8.42	8.28	8.55	–0.364
	Family labour	23.07	22.80	23.32	–0.30
	Expenditure	50,337.6	48,143.75	52,424.53	–0.71
	Seed (kg)	11.52	11.77	11.29	0.41
	Output (kg)	374.905	325.1333	422.2488	–2.4072**
	Yield (kg/acre)	112.932	103.8783	121.5447	–2.1708**
	Tech_eff. level	0.655	0.64	0.68	–2.57**
Income/revenue	Sales (Tsh)	164,281	142,282.1	185,207.3	–2.3705**
	Farm income	242,682.0	213,912.9	270,049.4	–1.3724*
	Non-farm inc.	14,306.3	15,202.56	13,453.76	0.5193
	T/income	256,989.0	229,115.5	283,503.2	–1.3183*
	H/h per capita	52,513.5	42,572.45	61,969.66	–2.5041**
	Agr.-asset Tsh	670,820	645,041	695,341.5	–0.6416
	Non-agr. (Tsh)	691,022	808,333.5	579,434.3	0.5160

			Per cent	
Credit access	Access to credit	4	4.1	3.9
Seed access	Better-seed	28.0	8.72	46.34
	Local seed	44.25	90.77	0
	Mixed-up	27.75	0.51	53.56
Extension	Ext-service	27	13.85	39.51
Education	No education	23.25	25.13	21.46
	Pr. education	74.0	73.33	74.63
	Sec. education	2.25	1.03	3.41
	Diploma	0.5	0.50	0.49
Gender	Male	79.25	75.9	82.4
	Female	20.75	24.10	17.56

Source: authors' fieldwork (2012).

Notes
*, ** and *** represent significance at 10%, 5% and 1% levels respectively.

Access to technology

Access to technology such as tractors improves land preparation, as well as dictating how much land can be cultivated in any given season. As indicated in Table 12.2, about 79 per cent of the households in the survey use ox ploughs. The extent of the usage of farming equipment shows no differences between contract and non-contract farming. This may explain the non-significance in difference of farm size between contract and non-contract farming. It also suggests that this factor does not contribute to the differences in productivity between the two groups.

Access to extension services

Extension services – particularly ensuring that farmers understand the agronomic practices of sunflower production – are critically important for obtaining the expected yields. As a crop, sunflower must strictly follow a specific cultivation calendar for different farming activities from sowing to harvesting. Proper spacing is also critical for yield results. The contract terms do not include the provision of extension services by the sunflower oil-processing companies because it was expected that extension services would be provided by the state through the office of the District Agricultural and Irrigation Cooperative (DAIC). The interaction of the DAIC and Firm A in the promotion of sunflower production through contract farming in Kongwa District has actually led to increased attention and training being offered to sunflower contract farmers by extension officers. As shown in Table 12.1, even though it is still below 50 per cent, access to extension services among contract farmers is nearly three times that of non-contract farmers (40 and 14 per cent respectively). In addition, some of the sunflower oil-processing companies have established demonstration plots to train smallholders and seed farms in the production of quality declared seed.

Table 12.2 Farm implements and inputs used

Type seed used	*Contract farmers*		*Non-contract farmers*	
	Frequency	*Percentage*	*Frequency*	*Percentage*
Improved seed variety	95	46	17	9
Both (improved and local)	110	54	1	1
Local seed variety			177	91
Total	205	100	195	100
Farm implements used				
Hand hoe	30	15	31	16
Ox plough	162	79	154	79
Tractor	13	6	10	5
Total	205	100	195	100

Source: authors' fieldwork (2012).

Access to market

The contract arrangements developed for farmers in Kongwa District had two terms regarding the market: first, that the contractor would establish a marketing centre in the villages, and second, the contractor would buy all the sunflower produced at the going price. According to in-depth interviews conducted in Iduo and Sugita villages in Kongwa District in December 2013, access to market did not seem to be problem for either contract or non-contract farmers. This is mainly because current sunflower seed production levels in the area and countrywide fall far short of the demand generated by the processing firms. This fact was corroborated by the processing firms, for example, the owner of a firm in Kibaigwa, who said that he used agents to buy even lower-grade sunflower seeds from non-contract farmers in order to use his installed capacity fully. The main area of contention as far as market is concerned has been the offered price, where contract farmers expect a premium. Hence farmers side-sell when offered a higher price than the contractor.

Access to information

Access to extension services entails access to different elements of agronomic information. In addition to increased public extension services to smallholders and the mobilisation seminars, the dissemination of agronomic information to smallholders through regular seminars was also planned to form part of the contractual system. The seminars were financially supported and organised by Firm A (see above), which played a key role in promoting contract farming and in 'linking' smallholders to sunflower oil-processing companies. However, the seminars were a one-off event conducted only at the piloting stage of the contract as a means to convince smallholders to participate in contract farming. Apparently, the information dissemination seminars were never repeated: according to the sunflower contract farmers who were interviewed, they only participated in one (mobilisation) seminar and never saw the people who conducted the seminars again.

This section has demonstrated the extent to which sunflower contract farming has empowered smallholder sunflower-growers economically by analysing what we have called the economic drivers in the contract, namely access to inputs, credit, markets, technology, information and extension services. Access to quality seeds appears to be the most important aspect of economic empowerment, with distinct differences in access and the use of quality seed between contract and non-contract farmers. The following section examines smallholders' political empowerment as a result of the sunflower contract farming arrangements.

Political empowerment

As defined in the model, smallholders are politically empowered by the contract if there is an institutional environment conducive to the economic decisions of the stakeholders involved, that is, the smallholders and processors. This section assesses the institutional environment of the contractual system according to the status of producer associations, the dispute settlement system and procedures, the capacity of the farmers to pursue such procedures and the trust-building initiatives introduced in parallel to the contractual system. The analysis focuses on the contractual system and does not make comparisons with non-contractual situations.

Existence of effective and accountable producers' associations

One of the components in the contractual framework was to promote and facilitate the formation of farmers' groups. This aspect is mentioned in the 'template' contract and was supposed to be initiated by the international NGO in collaboration with the processing firms. The purpose of forming farmers' groups is not elaborated in the contract, but the rationale seems to be that effective farmers' groups would enhance the bargaining power of otherwise weak smallholders in dealing with the processors. At the same time, the existence of a producers' association may reduce the transaction costs of having to deal with many individual farmers. It is probably for the latter reason that in some places the contractors (i.e. the processors) initiated and encouraged formation of farmers' groups. In Kibaigwa, for example, a processing firm attempted to mobilise farmers to pool their sunflower plots into larger blocks in order to increase efficiency in the provision of tractor services. Despite such isolated attempts to mobilise farmers into groups, all smallholders engaged in contract farming enter into individual contracts with the sunflower oil-processing companies. However, sunflower producer associations are generally absent in Kongwa District, possibly because of the newness of contracting farming arrangements in the area. The formation of producers' associations to enhance smallholders' bargaining power and interest is among the national strategic objectives in improving agriculture as elaborated in various policy documents (see, for example, URT, 2016), but this may need to be facilitated, as it does not happen spontaneously.

Existence of dispute settlement procedures

Contracts, whether formal with written regulations or informal relying on verbal agreements, are subject to dispute, as one or both parties may fail to comply with the agreed terms. Among the areas most prone to breach of contract is failure by the processors to deliver the right quantities of improved seed varieties to the contracted farmers. As emphasised above, only about half of the improved seed required by the smallholders is provided by the sunflower oil-processing companies. As a result, smallholders are forced to mix improved seeds with local seed varieties of lesser quality. For their part, smallholders breach contracts through side-selling. Due to the simplicity of the initially unwritten contract, it is difficult for either party to bring the other to court. Some of the farmers interviewed indicated that they needed written contracts to avoid conflicts. As a general rule, however, the Local Government Act No. 7 of 1982 (URT, 1982) empowers the village government and wards executives to enforce law and order within their areas. The village and ward councils have powers to enact by-laws enforceable within the village and ward respectively. In the absence of specific written conflict-resolution mechanisms in the contract farming arrangements, both the farmers and the processing firms expect any conflict arising between the contractors and the farmers at this level to be handled by the provision in the existing rules. Indeed, Firm A, an international NGO facilitating contract farming in the area, noted that, where ward and village executive officers were active and had taken steps against problems such as side-selling, contract farming has been quite successful. This was corroborated by farmers in Suguta and Iguo, who wanted to see greater involvement by village government in overseeing seed distribution, as well as in taking actions against farmers who side-sell and hence break trust with the processor.

Farmer's capacity in dispute settlement procedures

While the case for political empowerment incorporating written regulations and for dispute-resolution procedures and mechanism has been made, building the capacity of farmers in relation to dispute settlement procedures is probably the most difficult aspect of contract farming involving smallholders. As members of specific villages, farmers can participate in village assemblies, where they are free to raise any issue regarding their welfare. As provided for in the Local Government Act 1982, village councils are empowered to assist and advise villagers engaged in agriculture, forestry or any other economic activity. Whether farmers take advantage of such platforms in cases of dispute and exchange experiences and gain knowledge is not known, nor is it certain that the seminars conducted by the international NGO to promote contract farming also included civic education that would empower smallholder farmers. In any case, it is difficult for the smallholders to understand the complicated rules and procedures: as shown in Table 12.1, more than 95 per cent of the farmers, both within and outside the contractual scheme, have primary education at most, if any formal education at all.

Building of trust mechanisms

In the case of the sunflower contract farming in Kongwa District, building up trust between the parties as a motivation for the successful execution of a contract is probably more feasible than formal written contracts with no enforcement mechanism. Building up trust seems to have two dimensions. First, it is used as one of the sunflower oil-processing companies' strategies to obtain sufficient supplies of sunflower seed. This could be done with or without a contract. The strategies that have been deployed include offering cash advances to enable households to meet their cash requirements before the harvest for such things as school fees for children, medical fees or even purchases of food. One of the processors indicated a clear preference for entering into contracts with women (or with men – but in the presence of their wives!) because women were considered more trustworthy. Contracts with women were also believed to ensure that family welfare improves because women use cash advances and earned income for family welfare. Hence, being in a contractual arrangement with such possibilities provided through trust is therefore an aspect of political empowerment.

As Doney and Cannon (1997) observed, trust is considered to exist if one party believes that the other is honest. Smallholders do expect and trust processors to deliver the required quantities of seeds and other provisions promised in the contract. For instance, in Suguta and Iduo villages, smallholders indicated that they joined in the contractual arrangements because of the promised benefits, including the availability of quality seeds and an assured market. Processors trust that farmers would sell at least 50 per cent of the harvested crop in agreement with the contract terms.

However, only about a quarter of the contracted smallholders reported fulfilling their obligation to sell all their production to their respective contract buyers. Side-selling occurs when market conditions change so that contract farmers are able to sell to non-contract buyers at different and often higher prices than the average market price. In a way, therefore, processors are not competitive and cannot match the prices offered by other buyers. This constitutes an aspect of moral hazard whereby some farmers undertake side-selling their production to different buyers with whom they have no previous marketing agreements. Side-selling appears to be a major issue in explaining the failure of the existing contract system. For example, due to side-selling, Firm B decided to stop buying from contract farmers or even supporting them. Such behaviour indicates a lack of commitment and suggests the existence of basic problems with the present contracts. Overall, it does not appear that any political empowerment arises out of sunflower contract farming arrangements in Kongwa.

Reflections on the resulting economic outcome

The empowerment model adopted above suggests that the existence of economic and political empowerment would result in a number of economic outcomes. On average, contract farmers are not significantly different from non-contract farmers in terms of farm size and share of land allocated to sunflower production, which indicates that there is no pronounced specialisation in sunflower production among contract farmers. This section therefore looks at two main outcome indicators, yield and income, and asks whether there are differences between contract and non-contract farmers. Differences in income would tend to reflect differences in yield between the two groups, given that the sunflower contracts do not offer any price premiums. Contract farmers would obtain higher yields through improved productivity from using improved seeds, but they may also allocate more acreage to sunflower production due to the availability of seeds and hence obtain an even higher income.

While we observed an insignificant difference between contract and non-contract farmers in terms of land size and land allocated to sunflower production, contract farmers have significantly higher mean output and mean yields per acre. As shown in Table 12.1, while contract farmers have a mean output of 422 kg and mean yield per acre of 122 kg, mean output and mean yield per acre among non-contract farmers are only 325 kg and 104 kg respectively. This indicates that being involved in a contract gives some advantages that enable farmers to produce more output. As indicated above, these yields could probably be higher if farmers were provided with 100 per cent of the required improved seeds instead of the current 48 per cent provision. As shown in Table 12.1, about 53 per cent of the contract farmers mix local and improved seeds. Their weak political empowerment prevents them from taking action against the sunflower oil-processing companies that are supposed to provide the required volumes of improved seeds.

Availability of improved seeds in the sunflower contracts in Kongwa District does not seem to have prompted farmers to acquire more land or to allocate more land to sunflower production from family land holdings. Table 12.3 shows a mean of 3.3 acres among non-contract farmers compared to a mean of 3.7 acres among contract farmers, indicating a slight difference between contract and non-contract farms (significant at the 10 per cent level). Again this small difference could be due to the insufficient supply of improved seed to the contracted smallholders.

Table 12.1 shows that the mean technical efficiency level for all 400 farmers in the sample is 0.65 and that the mean technical efficiency among the contract farmers (0.68) is higher than that of the non-contract farmers (0.64); the difference is statistically significant at the 5 per cent level. The results show that, despite being higher, the efficiency levels of contract farmers are still far below their potential level of 1.0. However, contract

Table 12.3 Selected relevant aspects of sunflower production

Relevant variables in mean values	*Non-contract farmers*	*Contract farmers*	*Difference*	*t-test of difference*
Acreage	3.3 acres	3.7 acres	–0.4	–1.44*
Share of sunflower land	0.47	0.49	–0.02	–1.06
Yield per acre	103.9 kg	121.5 kg	–17.7	–2.17**
Mean output	325.1 kg	422.2 kg	–97.1	–2.41**

Source: authors' fieldwork (2012).

Notes
***, ** and * represent significance at 1%, 5% and 10% levels respectively.

farmers have significantly higher sales revenue, total non-farm income and household income per capita compared to non-contract farmers, which indicates that, by being involved in contracts, farmers gain certain advantages over non-contract farmers that enable them to improve their production, productivity and income. Mean income for contract farmers for the season in question was 270,049 Tsh as opposed to 213,912 Tsh for non-contract farmers, with the results being statistically significant at 5 per cent. Notably, the differences in the economic outcomes between contract and non-contract farmers arise only from economic empowerment.

Conclusion

This chapter has investigated whether contract farming empowers smallholders by comparing contract and non-contract sunflower farmers in Kongwa District, central Tanzania. The examination was based on two forms of empowerment, economic and political, and used an analytical model developed from the World Bank's definition of the concept and a conceptual framework adopted from Utting (2012).

Using the economic drivers of empowerment with reference to the particular production assistance accessible only to farmers who participate in the contract arrangements, it is clear that contract farming does empower smallholder farmers economically. This result is substantiated by the economic outcomes in terms of higher mean outputs and incomes for contract farmers as compared to non-contract farmers. On the other hand, non-fulfilment of some of the contract terms, in particular the inadequate supply of improved seeds, resulted in lower than potential economic outcomes in terms of productivity, incomes and efficiency levels.

Concerning political empowerment, the study showed virtually no effect of the contract arrangements. Admittedly, the selection of farmers and hence participation in the contract was open to any farmer motivated by the seminars held by the contract farming facilitators. But while the contract arrangements made provision for the setting up of farmers' associations to represent farmers' interests and increase their negotiation power, such associations never materialised, perhaps due to the absence of organs such as a cooperative society that could organise farmers. As a result, each contract farmer dealt with the processor alone. Hence institutional support in assisting smallholders to form associations still remains pivotal as a means to empower them.

The attempt by some processors to build up and gain the trust of individual farmers through the provision of assistance outside the contract could be seen as a double-edged sword. On the one hand, it strengthened the farmers by enabling them to meet their daily requirements by providing them with cash credit. On the other hand, it weakened them politically by locking them into the contract, even when they were not happy with some of its terms, for example, having to sell at what they perceived to be a lower market price. Therefore, it is not surprising that, despite the attempts to build up trust, some contract farmers decided to side-sell. While this could be seen as farmers breaking the contract terms, it could also indicate that the smallholders attempt to empower themselves against what they perceive to be an 'unfair price'.

The sunflower contract farming dispute-resolution mechanism was left to the existing local governance structures. Given the simplicity and newness of contract farming in the district, it seems reasonable to use the existing dispute-resolution mechanism from the perspective of both cost and effectiveness. Where the village executive officers were active in enforcing disputes using the available laws and by-laws, the contracts actually performed better and side-selling was minimised. In any case, the existence of a dispute-resolution mechanism by itself may not empower farmers if it cannot be enforced.

In order to improve the 'resulting economic outcome' and rural development, the intertwined relationship between economic empowerment and political empowerment needs to be acknowledged by international donor agencies and policy-makers in Tanzania. Farmers' associations have not developed as envisaged in the sunflower production areas of Singida and Dodoma in central Tanzania as they have in the case of privatised tobacco schemes in Tabora and sugar schemes in Kilombero. We would argue that as a *sine qua non* political empowerment entails deliberate efforts and institutional support to facilitate associations and farmer groups that are capable of bargaining for key economic drivers, including extension services and the delivery of improved seeds and other inputs.

References

Bandura, A. (1997). *Self-efficacy: The Exercise of Control.* New York: Freeman.

Bellemare, M. F. (2012). As You Sow, So Shall You Reap: The Welfare Impacts of Contract Farming. *World Development*, 40(7), 1418–1434.

Cai, J., Setboosarng, S. and Leung, P. (2014). Contract Farming and Poverty Reduction: The Case of Organic Rice Contract Farming in Thailand. In S. Setboosarng and P. Leung (eds), *Making Globalization Work through Contract Farming* (pp. 71–108). Mandaluyong, Manila: Asian Development Bank.

Conger, J. A. and Kanunga, R. (1988). The Empowerment Process: Integrating Theory and Practice. *Academy of Management Review*, 13(3), 471–482.

De Schutter, O. (2012). *Contract Farming Can Work.* Capacity.Org. A Gateway for Capacity Development. Geneva: United Nations Research Institute for Social Development.

Doney, P. and Cannon, J. P. (1997). An Examination of the Nature of Trust in Buyer–Seller Relationships. *Journal of Marketing*, 61(2), 35–51.

FAO (2011). *The State of Food and Agriculture: Women in Agriculture; Closing the Gender Gap for Development.* Rome: FAO.

Fold, N. and Larsen, M. N. (2011). Upgrading of Smallholder Agro-food Production in Africa: The Role of Lead Firm Strategies and New Markets. *International Journal of Technological Learning, Innovation and Development*, 4(1/2/3), 39–66.

Gereffi, G. (1999). International Trade and Industrial Upgrading in the Apparel Commodity Chain. *Journal of International Economics*, 48, 37–70.

Gibbon, P. (2001). Agro-Commodity Chains: An Introduction. *IDS Bulletin*, 3(32), 60–68.

Ibrahim, N. H. and Allen, D. (2012). Information Sharing and Trust during Major Incidents: Findings from the Oil Industry. *Journal of American Society for Information Science and Technology*, 63(10), 1916–1928.

Key, N. and Runsten, D. (1999). Contract Farming Small Holders, and Rural Development in Latin America: The Organization of Agro Processing Firms and the Scale of Out Grower Production. *World Development*, 227(2), 381–401.

Mpeta, D. F. (2015). Effects of Contract Farming on Production and Income of Sunflower Farmers in Kongwa Districts in Central Agricultural Zone in Tanzania. Unpublished PhD Dissertation. Mzumbe University.

Mushumbusi, P. K. and Kratzer, J. (2013). Empowering Women through Microfinance: Evidence from Tanzania. *ACRN Journal of Entrepreneurship Perspective*, 2(1), 31–59.

Narayan, D. (2005). Conceptual Framework and Methodological Challenges. In D. Narayan (ed.), *Measuring Empowerment: Cross Disciplinary Perspectives* (pp. 3–38). Washington, DC: World Bank.

Petesch, P., Smulovitz, C. and Walton, M. (2005). Evaluating Empowerment: A Framework and Cases from Latin America. In D. Narayan (ed.), *Measuring Empowerment: Cross-Disciplinary Perspectives* (pp. 39–67). Washington, DC: World Bank.

Prato, B. (2013). *Supporting Poor Rural People's Empowerment through Policy Solutions for Natural Resources Management and Agriculture*. Paper prepared for the UN Expert Group Meeting (EGM) on Empowerment, 10–11 September 2013, United Nations, New York.

URT (1982). *Local Government Act No. 7 of 1982*. United Republic of Tanzania.

URT (2016). *National Five Year Development Plan, 2016/17–2020/21*. United Republic of Tanzania.

Utting, P. (2012). *The Challenge of Political Empowerment*. Capacity.Org. A Gateway for Capacity Development. Geneva: United Nations Research Institute for Social Development (UNRISD).

World Bank (2002). *The World Bank: Empowerment Sourcebook*. Washington, DC: World Bank.

Zakaria, Y. (1999). *Entrepreneurial Ethics and Trust: Cultural Foundations and Networks in the Nigeria Plastic Industry*. Brookfield: Ashgate.

13 Embedding the global tobacco value chain in social and environmental concerns

Contract farming and corporate social responsibility projects in the Tanzanian tobacco sector

Marianne Nylandsted Larsen and Jonas Gillett

Introduction

The tobacco industry is one of the most contested global industries, with strong public and private anti-smoking pressure groups and leading international opponents such as the World Health Organization (WHO) questioning the legitimacy of the industry's activities. After several 'dark decades' of denying the health risks associated with cigarette consumption, the tobacco industry had to acknowledge the lethal and addictive character of smoking during the 1990s (Palazzo and Richter, 2005; Gonzalez *et al.*, 2011). The tobacco industry was forced to introduce health warnings on cigarette packs, while several countries (including the United States and countries in Europe) outlawed broadcast tobacco advertising in an attempt to reduce cigarette consumption. In the early 2000s, however, the major international cigarette branders became increasingly exposed to allegations of negative labour practices and the damaging environmental implications of tobacco production in developing countries. Public, high-profile media campaigns questioned the sustainability of the tobacco industry and its exploitive character. As a response to criticisms regarding persistent negative labour practices – notably child labour engaged in tobacco-related work on farms – and the vast deforestation of woodlands to provide tobacco-growing areas, the major international cigarette companies integrated these concerns into their corporate social responsibility (CSR) programmes to mitigate the risk of damage to their brands. In addition, their exposure to public awareness spurred a number of financial initiatives and strategic alliances between the cigarette companies and their tobacco leaf suppliers in order to demonstrate that ethical and environmental concerns are being addressed in the tobacco-producing countries. During the 2000s, the big cigarette companies adapted and 'greened' their value chain practices by integrating environmental and labour considerations into them in an attempt to position the industry as socially and environmentally responsible (Otañez and Glantz, 2011: 1). As a result, an increasing number of performance requirements have been transferred from cigarette branders to their leaf suppliers (tobacco-processing companies) in tobacco-growing countries.

This chapter explores the recent emphasis on social and environmental concerns in contract tobacco-farming schemes in Tanzania. The empirical focus is on the tobacco value chain, defined as a sequence of value-added activities from tobacco production at the farm level via trade in tobacco blends on the world market to cigarette manufacturing at the cigarette company level. The tobacco value chain consists technically of three basic nodes: the cultivation of tobacco leaves, the processing of leaves and the manufacturing of cigarettes. Tanzania is sub-Saharan Africa's third biggest tobacco producer after Zimbabwe and Malawi, and tobacco production provides an important livelihood for an estimated 140,000 Tanzanian smallholders, mainly located in Tabora Region, south-west Tanzania. Processing of leaves is carried out by two tobacco-processing companies located in Morogoro, and more than 85 per cent of all tobacco produced in the country is exported to a handful of large international cigarette companies with strong brands, including British American Tobacco, Philip Morris and Japan Tobacco International. The chapter argues that the cigarette companies' recent engagement in CSR and the promulgation of strategies in relation to social and environmental concerns are influencing the governance of the global tobacco value chain, the organisation of the contractual system in Tanzania and the content of the contract. Accordingly, the responsibility for ensuring compliance with key CSR requirements (notably agricultural and labour practices) has been transferred to the tobacco-processing node of the value chain and is now monitored by the branded cigarette manufacturers. In order to ensure compliance with crucial CSR requirements, the Tanzanian tobacco-processing companies tightly coordinate the contract farming schemes and inspect and monitor farm production processes (notably related to husbandry and labour practices) through a joint-venture subsidiary, the Association of Tanzanian Tobacco Traders. Interventions in the production process are consolidated through the contractual agreement between the tobacco-processing companies and the primary cooperative societies, which stipulates the respective parties' obligations to ensure compliance with good agricultural and labour practices. At the same time, the tobacco-processing companies are increasingly making voluntary contributions to local communities as part of their CSR programmes, with co-financing by the international cigarette companies. The ability to finance development projects and allocate resources both among and within communities to individual farmers complements the tobacco companies' strategic focus on obtaining a steady supply of tobacco leaves by rewarding loyal primary cooperative societies, while addressing compliance with crucial CSR requirements through the elimination of child labour and good agricultural practices.

The chapter is organised into six main sections. Following this introduction, the next section engages with recent developments in governance mechanisms employed by the major international cigarette manufacturers – the so-called lead firms in the global tobacco value chain – by focusing on strategic alliances, multi-stakeholder initiatives and the implementation of CSR strategies to combat criticism and secure brand reputation. The third section moves to the Tanzanian context. First, the contractual schemes in the tobacco sector are briefly outlined, followed by an analysis of recent efforts and initiatives by the tobacco-processing companies to comply with CSR requirements. The fourth section explores CSR projects directed towards socio-economic development activities at the community and/or individual farm levels, while the fifth section provides some broader reflections on the potential and limitations of private-sector financed local development projects driven by the concerns of international cigarette companies. Some concluding remarks are provided in the final section.[1]

The battlefield: public allegations and engagement with CSR programmes

With the closure of the 'dark decades' of the tobacco industry, a new set of allegations surfaced in public debates, becoming a threat to their very existence.[2] Reports by WHO, labour and human rights activists, environmental groups and anti-smoking NGOs started highlighting the extensive use of child labour, deforestation, soil depletion and erosion, and pesticide and nicotine poisoning of workers, particularly in developing countries (Otañez and Glantz, 2011; Jamil, 2010). For example, at the end of the 1990s high-profile documentation of child labour in the Malawian tobacco sector by the International Labour Organization (ILO), the International Confederation of Free Trade Unions and IUF[3] threatened the reputations of the big international cigarette manufacturers at a time when public (consumer) awareness and concerns over the social and environmental impacts on production and trade were growing rapidly (Otañez *et al.*, 2006). As the tobacco product at the retail level is highly susceptible to brand visibility, the cigarette manufacturers spurred various individual and collaborative initiatives in order to mitigate brand damage and control vital issues that challenge the legitimacy of the tobacco value chain.

In 2000, British American Tobacco, which had an estimated 11 per cent of the global cigarette market, signed a memorandum of cooperation with the International Tobacco Growers Association (covering the major cigarette companies and tobacco leaf suppliers) and the IUF. The three organisations agreed to work jointly to combat child labour in tobacco-growing countries. Organisationally this was followed up by the establishment of the Eliminating Child Labour in Tobacco Growing (ECLT) foundation in 2001. The membership base of the ECLT foundation include all the major international cigarette companies and major tobacco leaf-processing companies such as Dimon, Standard Commercial Corporation and Universal Leaf Tobacco (ECLT, 2013a), and it has become one of the most important organisational structures coordinating and supporting programmes to reduce tobacco-related child labour in tobacco-growing countries. Since 2002, the ECLT Foundation has financed 23 projects to combat child labour in seven tobacco-growing countries in collaboration with local NGOs, farmers' associations, trade unions, tobacco companies and government ministries. These projects generally include awareness campaigns at the national, district and local levels on the hazards of child labour in tobacco-growing activities. In addition, financial support is provided for school and educational activities (the construction or rehabilitation of schools), as well as directly to farmers to improve their livelihoods and eliminate their use of on-farm child labour (ECLT, 2013a). In addition, the ECLT foundation provides an important advocacy platform for infusing the global tobacco value chain with some kind of ethnical accountability: it has recently launched a 'Pledge of Commitment' by ECLT foundation board members' uniting stakeholders throughout the entire tobacco value chain to eradicate child labour. Collaboration with 'chain-external' partners (e.g. international NGOs, multilateral organisations such as the ILO) has been rare as a result of resistance to or the fear of becoming associated with the 'addictive and lethal character' of the products the cigarette industry sells. The international NGO Save the Children, however, has recently joined the ECLT foundation board as an adviser, and both the ILO and Save the Children Switzerland were consulted during the drawing up of the Pledge of Commitment (ECLT, 2014, 2015).

Besides the formalisation of the ECTL foundation, the big branders have initiated CSR programmes and developed mechanisms to control and monitor their tobacco leaf suppliers in order to address and legitimise the image of tobacco farming as socially and environmentally responsible. In the late 1990s, British American Tobacco pioneered the tobacco industry's commitment to CSR by publishing a report on the company's environmental and agricultural performance, while Philip Morris launched its 'Tobacco Identity Preservation Program' in 2000 to track tobacco from 'farm to cigarette'. By their very nature, the CSR programmes and the promulgation of strategies in relation to social and environmental concerns go beyond compliance with society's legal and basic moral rules and entail connotations of 'giving something back' to society. Accordingly, 'good agricultural practices' and 'good labour practices' now feature prominently in the cigarette branders' CSR programmes (see, for example, Alliance One, 2016).[4]

Generally, the CSR programmes contain guidelines, rating criteria, action plans and auditing methods for their supply chain participants in order to increase control over the production process. Philip Morris's CSR programme, 'Good Agricultural Program: Guidelines and Assessment', set out performance requirements for its tobacco leaf suppliers and a method for measuring suppliers' performance in six areas: 'mission and values (including addressing child labour), variety management and integrity, crop management, integrated pest management, sustainability (including participation in reforestation projects) and product integrity' (Otañez and Glantz, 2011: 2).

British American Tobacco's 'Social Responsibility in Tobacco Production: Road Maps' outlines the company's efforts to integrate supply chain activities with CSR schemes. It covers all major areas of tobacco production in the value chain, such as integrated crop management, tobacco processing, workplace environmental health and safety, and socio-economic factors like child labour, farmers' living standards and capacity-building.

In most cases, the big cigarette manufacturers monitor their tobacco leaf suppliers' social and environmental performances through check-list-based audits combined with the tobacco leaf suppliers' self-assessments of their own performances. This is supported by a scoring system related to the key pillars of the cigarette companies' CSR strategies. In addition, British American Tobacco, Philip Morris and other branded cigarette manufacturers increasingly use independent auditing companies (e.g. Leaf Technology Consultancy) to review their tobacco leaf suppliers (Palazzo and Richter, 2005; Otañez and Glantz, 2011).

Efforts to secure brand reputation and the recent implementation of CSR strategies by the handful of large branded cigarette manufacturers that dominate the global cigarette market have resulted in a tighter form of governance. They have developed mechanisms to coordinate and monitor their supply chains in order to keep records of the origin of the tobacco they use and also of the conditions under which it has been produced. The cigarette manufacturers allocate responsibility for conformity with good agricultural and labour practices to their suppliers, and a number of performance requirements are imposed on tobacco-processing companies in tobacco-growing counties. In this way, the big cigarette manufacturers define the key terms of participation in the tobacco value chain directly for their tobacco leaf suppliers and indirectly for other actors further upstream in the chain. The following sections analyse the relationship between contract farming and CSR concerns in the tobacco value chain originating in Tanzania. The analysis shows how contract farming, the organisation of the contractual system and local socio-economic development initiatives provide insurance against critical voices, both locally and internationally (Jamil and Neville, 2011).

Infusing social and environmental concerns through contract farming in Tanzania

The tobacco industry in Tanzania was liberalised in the mid-1990s, and the former parastatal processing plants were privatised in 1997 (Mitchell and Baffes, 2002). One of the processing plants (located in Morogoro) was sold to Universal Leaf Tobacco, one of the world's leading leaf tobacco merchants and processors that handles between 35 per cent and 45 per cent of the annual production of flue-cured and burley tobacco in Africa (Universal Leaf Tobacco, 2016). The Tanzanian subsidiary operates under the name of the Tanzania Leaf Tobacco Company (TLTC) and supplies tobacco blends to several of the large multinational cigarette manufacturers, including Philip

Morris, whose Marlboro and Benson & Hedges are its most famous brands (Philip Morris, 2016).[5] In addition, Dimon – later renamed Alliance One as a result of the merger of Dimon Incorporated and Standard Commercial Corporation – invested in a new factory located in Morogoro. In addition to sale of tobacco blends to Philip Morris, Alliance One has long standing trading links to British American Tobacco (brands: Pall Mall, Lucky Strike and Kent) and Japan Tobacco International, which owns the brands Camel, Winston and Mild Seven (Alliance One International, 2016; Japan Tobacco International, 2016).

Following the dismantling of the former parastatal marketing channel, the primary cooperative societies (PCSs) agreed to sell tobacco leaves under contract to the two private tobacco-processing companies. Seeds, fertiliser and chemicals were supplied on credit to the PCSs, and the costs of inputs were deducted when the smallholders sold their tobacco produce to the company through their respective PCS. Side-selling became a major problem a few seasons after the input credit schemes were launched, as smallholders took input loans from one company and sold tobacco to another company or a local trader in order to avoid repaying loans. In order to prevent side-selling but still retain their ability to supply essential inputs to smallholders on credit and provide extension services, the two tobacco companies formed a wholly owned subsidiary, the Association of Tanzanian Tobacco Traders (ATTT). The strategic alliance between the two tobacco-processing companies was initially established as a result of side-selling practices by smallholders, combined with a need to increase consistency and achieve superior crop quality. Hence, the main activities of the ATTT were related to improvements to crop quality and yields as a means to earn higher prices on the world market and included training in new on-farm husbandry practices related to soil preparation, fertiliser and pesticide application and processing practices (e.g. curing barns), in addition to the coordination and recovery of input credits at the PCS level (Mitchell and Baffes, 2002).

During the 2000s, ATTT embarked on a modification of its internal operating procedures in order to comply with the industry's CSR programmes, and it became the most important organisational entity coordinating and monitoring smallholder compliance with crucial CSR requirements (see Table 13.1). In 2009, the input distribution system was transferred from ATTT to the cooperative unions (WETCU in the case of Tabora) and their umbrella organisation, the Tanzania Cooperative Apex, as part of the government's efforts to 're-empower' the cooperative system in Tanzania. Hence, the two tobacco-processing companies were excluded from involvement in the provision of inputs to smallholders, and ATTT's activities became solely focused on providing extension services and monitoring tobacco cultivation in the main tobacco-producing region (interviews with tobacco processing companies, Morogoro, July 2013, and ATTT, Tabora and Urambo, July 2013).[6] According to ATTT, 90 per cent of their time and financial resources are currently centred on successfully implementing the three main pillars of

Table 13.1 Overview of the CSR programmes in the Tanzanian tobacco contract scheme

CSR pillars	*Coordination*	*Target group(s)*	*Provider of financial resources*	*Specific aims*
Labour practices, including elimination of child labour	Contract content Monitored by ATTT	All PCSs and farmers	Tobacco processing companies Total LandCare and ECLT Foundation	Brand protection through increased compliance
Good agricultural practices	Contract content Monitored by ATTT	All PCSs and farmers	Tobacco processing companies	Brand protection through increased compliance
Reforestation projects	Contract content Monitored by ATTT	All PCSs and farmers	Tobacco processing companies Total LandCare and ECLT Foundation	Brand protection through increased compliance Ensure reproduction of fuel wood for curing
Social responsibility projects	Monitored by tobacco processing companies	Selected PCSs	Tobacco processing companies International cigarette companies	Brand protection through increased compliance PCSs to (re)sign contracts
Farmer packages	Monitored by ATTT	Selected farmers	Tobacco processing companies International cigarette companies	Brand protection through increased compliance. Increase farmer returns/productivity PCSs to (re)sign contracts

Source: authors' fieldwork (2013, 2014).

the CSR programmes: (1) good agricultural practices; (2) safe labour practices, including elimination of child labour; and (3) reforestation projects (interview with ATTT, Tabora, March 2014). ATTT employs about 600 permanent staff, of whom 300 are directly involved in in-field training and the supervision of farmers.

The geographical area covering tobacco production in Urambo and Sikonge Districts is divided into 26 'units', and through carefully planned training circles, so-called field technicians regularly monitor and disseminate information to farmers. The training of farmers is mainly conducted using role models: strongly performing contract farmers are selected for training, and they in turn are responsible for relaying information to other group members. The field staff are instructed to visit two contract farmers a day (two weeks for one training/monitoring circle). In addition, they regularly inspect and supervise individual farmers in on-farm husbandry practices (interviews with ATTT, Urambo, July 2013 and tobacco-processing company, Morogoro, July 2013).

In the contract farming scheme, the PCSs constitute the main entity between the tobacco farmers and the tobacco-processing companies, and as such, justifying the conditions connected with the organisation of tobacco production falls to them. The tightly coordinated contract scheme and interventions in the production processes are consolidated through the contractual agreement between the tobacco-processing companies, the PCSs and WETCU. The contract contains details on the division of labour and the respective parties' responsibilities in relation to input supply, selling and purchasing obligations. Further, the contract stipulates the obligations of the PCSs to ensure that their members act in accordance with good agricultural practices and comply with reforestation and labour-practice requirements. The phrase 'good agricultural practices' refers not only to rules regarding the application of agrochemicals (including fertilisers), but also to the provision of sufficient barns for curing tobacco leaves, though the contract does not make detailed specifications. Tobacco requires substantial amounts of wood for the curing process, and in order to combat deforestation and environmental concerns (see above), the contract specifies that each PCS member is obliged to grow a minimum of 50 trees for each bag of fertiliser supplied on credit, which represents the equivalent of 500 trees per hectare of flue-cured tobacco. Further, for curing farmers are required to use only wood from planted forest or supplied by the government in order to preserve the native forest.[7] The issue of labour practices is addressed through various requirements, for example, using protection gear when applying toxic pesticides or handling green tobacco (to avoid contact with the skin), and PCSs are required to ensure that their member farmers 'in particular, do not engage or involve children in any tobacco related business, in contravention of applicable laws and to the extent that it might compromise the health, education or wellbeing of the child' (Tobacco Buying Agreement, n.d.: 6). In order to ensure compliance with CSR requirements, ATTT staff rigorously monitor and train farmers and PCSs. In cases of non-compliance by PCS members (e.g. evidence of child labour), ATTT reports its observations to the companies concerned, with termination of the contract as a likely consequence.

Besides efforts to secure brand reputation through the contractual systems, the big cigarette branders – British American Tobacco, Philip Morris, Japan Tobacco International and Imperial Tobacco – fund child labour prevention and deforestation projects in Tanzania through Total LandCare and the ECLT foundation (see Table 13.1). The ECLT foundation announced its first child labour-elimination project in Tanzania in 2004, and the latest project (2011–15, budgeted at US$4.7 million) was coordinated and implemented by Winrock International, the Tanzania Association of Women Leaders in Agriculture and the Environment and the Tabora Development Foundation Trust (ECLT, 2013b). One of the key components in this project is to organise so-called child labour committees and provide training and procedures for the systematic monitoring of child labour in tobacco-related activities. Another key component of the project is advocacy for mainstreaming child-labour issues into district development plans and budgets.

CSR projects at the community level

In addition to embedding social and environmental concerns in the contract farming schemes, the Tanzanian tobacco-processing companies are increasingly making voluntary contributions to local communities as part of their CSR programmes. These projects are co-financed by the major cigarette branders, and since the late 2000s the number of projects has increased substantially. Also, financial support to social and economic development activities at the community and farmer levels has become more formalised through so-called social responsibility projects and farmer packages. These two main modes of conveying socio-economic development activities at the community level are considered in turn in this section.

Social responsibility projects at the community level

Since 2009, the two tobacco-processing companies have systematically supported social and economic development activities at the village level. The social responsibility projects (SRPs) support community-level initiatives within three main areas: education, health and the environment. As in the case of the farmer packages, these CSR projects are co-funded by the tobacco-processing companies and cigarette branders, though the financial budget for SRP initiatives varies from year to year, as well as between the two tobacco-processing companies (interview with tobacco-processing companies, Morogoro, March 2014). Financial support to local communities is only provided in the form of in-kind donations (e.g. building materials) and only awarded to villages that have demonstrated they have financial means and have been able to initiate the project before an application is approved by the tobacco company. In other words, in order to become eligible for funding, the PSC needs to demonstrate financial capacity and planning capabilities, while the project, once approved for financial support, is monitored through regular visits by the particular tobacco company's area representative. As shown in Table 13.2, the majority of the SRP projects received by the PCSs that were interviewed involves the construction of schools (or the rehabilitation of classrooms) and dispensaries, in addition to improvements in access to water and health care.

The tobacco-processing companies sign identical contracts with the PCSs for a three-year period. The 'linking' of PCSs to the tobacco companies or the re-signing of an established contract is configured through yearly scheduled 'touring around campaigns', where promises of financial support to loyal PSCs, formalised through the SRPs, form part of the tobacco companies' competitive strategies in terms of securing sufficient tobacco volumes. ATTT is not directly involved in the community-based SRP initiatives (see Table 13.1), but it reports regularly on compliance and non-compliance to the tobacco-processing companies, which provides important information for the evaluation of applications eligible for financial support at the quarterly assessment meetings at the headquarters in Morogoro.

By financing what seem to be quite ordinary local development projects, the ability to allocate resources among communities complements the tobacco companies' strategic focus on the consistent supply of tobacco leaves (rewarding loyal PSCs) while at the same time addressing the issue of compliance with crucial CSR requirements. Considerable efforts are made to forge close relationships and reward the three-year contracts with PSCs that have demonstrated their capabilities in terms of high and steady seasonal production, for which they are rewarded in the form of substantial community assistance (for example, PCS7 and PCS11, Table 13.2). Financial support through SRP initiatives is, to some extent, also perceived as a means to increase the capacity to address vital CSR strategies at the community level. As an employee interviewed at ATTT emphasised:

> If the schools are run down, there are no books, no chairs, and no desks and even worse, in some cases no school at all; how can anyone expects the kids to go to school and not to the field?
>
> (Interview with ATTT, Morogoro, March 2014)

By having one of the SRP's components focused on in-kind donations for improvements to school and educational facilities in particular, the industry can facilitate access to better-functioning primary schools at the community level and hence provide the foundation for 'a cultural change among the farmers' (interview with ATTT, Morogoro, March 2014). PCS loyalty in terms of securing determined tobacco volumes at the time of crop sales to the tobacco companies remains crucial, but the decision to finance school and educational facilities is also based on observed cases of child labour in

Table 13.2 PCSs' characteristics and kinds of SRPs received by PCSs

PCSs	*Year of establishment*	*No. of farmers*	*Output (kg)*	*SRP*	*Year of approval*	*Amount (million Tsh)*	*Processing company*
PCS1	1987	355	428,000	Construction of classroom	2010	3	Alliance One
PCS2	1988	400	413,956	Construction of dispensary	2010	50	Alliance One
PCS3	1988	126	186,000	Desks and tables for school, rehabilitation of PSCs office	2013	4.1	Alliance One
PCS4	2003	228	446,000	Desks for school	2011	3.5	Alliance One
PCS5	2007	60	85,167	Rehabilitation of classroom	2009	3	Alliance One
PCS6	1988	560	460,000	Books for school	2011	3.8	Alliance One
PCS7	1988	667	778,000	Construction of dispensary	2014	40	TLTC
PCS8	1998	250	280,000	Desks for school, rehabilitation of teacher's house	2009	2.4	TLTC
PCS9	2009	203	150,000	Construction of dispensary	2012	30	TLTC
PCS10	1975	360	300,000	Desks for school	2010	8	TLTC
PCS11	1984	800	1,050,000	Construction of storage facilities, wells, tractor, rehabilitation of PCS office	(2005) 2011	Unknown	TLTC

Source: authors' fieldwork (2014).

tobacco-related activities. Hence, in some cases, despite non-compliant behaviour, some PCSs are successful in applying for financial support. The reason behind this somewhat paradoxical way of providing CSR-related projects (financial support to PCSs based on observed 'wrong' labour practices, notably abusive child labour) is obviously to improve the ability of communities to conform to labour requirements.

Taken together, the SRP initiatives have direct development impacts by contributing to the construction and maintenance of, for example, local public goods and services. In general, there is no usage charge placed on the projects once they have been constructed, and most of the supported development projects effectively become part of the state (village) infrastructure that all community members normally benefit from – in other words, the benefits are not restricted to the PCS leadership or certain farmers. In addition, the projects have been credited with bringing several social benefits, including community mobilisation for development projects.

Farmer packages

The farmer packages are the latest component of the CSR programmes. They were initiated in 2012 by both tobacco-processing companies with the aim of supporting well-performing farmers and PCSs, and thus indirectly these 'packages' aim at providing incentives to other farmers (and PCSs) to 'copy' the good husbandry and labour practices adopted by the well-performing farmers. Farmer packages are directed at a few progressive and well-performing farmers within the contractual schemes. The criteria for selecting farmers for in-kind support consists of two layers: first, loyal PCSs are identified, and second, in collaboration with the chairman of the selected PCS, three to five of the best-performing farmers who can demonstrate their compliance with labour and agricultural requirements (including reforestation) are selected for financial support. ATTT staff assist either Alliance One or TLTC with the selection of PSCs and individual farmers based on excellent production results, consistent compliance with required labour and good agricultural practices, and their assessed ability to utilise the equipment provided to them efficiently. As indicated in Table 13.3, the specific products provided to individual farmers or PCSs include both equipment directly related to improvements in tobacco production (clips for handling tobacco leaves, barns, oxen, trolleys) and water projects such as the provision of shallow boreholes and treadle pumps. Apart from oxen, 'farmer package products' are provided in-kind to individual farmers. A pair of oxen and associated training is provided either as a loan with a repayment clause or in-kind for a limited period of one year. In the latter case, it is expected that increasing productivity as a result of improved land preparation will provide the farmer with the financial means to purchase a pair of oxen after the 'trial period'. However, according to Otañez and Glantz (2011: 9), the cost of a pair of oxen and associated training is US$320, 'which is beyond the reach' of most tobacco farmers in Tanzania.

Table 13.3 Farmer package products and selection criteria

Products	*Selection criteria*
Clips	Based on PCS and farmer compliance
Improved barns	Productive PCSs showing good leadership
Oxen	Loyal PCSs
Trolleys	Productive and competent farmers
Water filters	Farmers showing progressive behaviour
Wells including treadle pumps	PCSs

Source: authors' fieldwork (2014).

More detailed selection criteria for each farmer package product are given in Table 13.3. The choice of specific products is based on the degree to which it will improve the farmer's productivity. One of the commonest farmer package products provided in-kind is the construction of a new barn and/or optimising the use of the available barn capacity. Tobacco leaves are cured in heated barns for about a week, and the quality of the curing barn influences the quantity of fuel wood required for the curing and the quality of the cured tobacco leaves (Geist *et al.*, 2009). Agricultural equipment for handling tobacco or undertaking better land preparation (draught animals) enables these farmers to achieve higher yields and levels of production, as well as achieving more remunerative grades and higher tobacco leaf prices. Hence, the provision of farmer packages by the tobacco-processing companies might greatly improve incomes and the level of assets held by the selected farming households. In addition to financial support to individual farmers within the PSCs, improvements to water conditions (shallow boreholes and treadle pumps) are typically directed towards very loyal PSCs only, but they are of general benefit to the whole community.

CSR programmes and local socio-economic development

The role of the tobacco-processing companies in shaping processes of socio-economic change at the tobacco community level has increased profoundly in recent years. When CSR programmes 'touch down' in local settings, however, they inevitably interact with local patterns of socio-economic development and local relations of power. As such, therefore, the SRP projects and farmer packages should not be viewed as neutral development tools. The tobacco-processing companies' CSR projects provide new constellations of economic dynamics and opportunities for both PCSs and farmers. While the longer-term impacts on local socio-economic development remain to be seen, the conditions associated with the recent development dynamics might point to increasing inequality and socio-economic differentiation between the PCSs in the tobacco-growing area, as well as among tobacco farmers within

PCSs themselves. Furthermore, tensions can easily arise between different local development agencies (e.g. village government vis-à-vis PCS leadership), who represent or reflect different development interests at the community level.

Indications of increasing socio-economic differentiation between the PCSs in the tobacco-growing area can be detected by sketching out the parameters for obtaining SRP projects: the decision-making and planning processes, allied to applications for financial support, are rooted at the PCS level, as they are the key entry points through which the tobacco-processing companies allocate resources. In order to become a recipient through an SRP project, the PCS leadership has to demonstrate decisive competences to plan and raise the means to co-finance the project. Hence, the realisation of development opportunities provided by the tobacco-processing companies depends critically on the existence of PCSs with leadership qualifications, capacity and capabilities and the financial means to move development in a certain direction. Furthermore, access to resources allocated by the tobacco-processing companies are driven by the concerns of the international cigarette companies, and as mentioned above, SRP projects are used strategically by the tobacco-processing companies to reward PCSs that can demonstrate compliant production processes and loyalty. This may eventually result in imbalances and social inequalities between those PCSs that are of 'value' to the tobacco companies in terms of being able to offer reliable supply combined with a high degree of compliance with the CSR requirements and those PCSs that are of limited interest to the tobacco companies.

Within those PCSs of continued interest or value to the tobacco-processing companies, the ongoing process of financing 'ordinary' local development projects (e.g. schools) might cause tensions between various local development agencies. The SRP projects tend to place the PCSs in a position where they 'replace' the state (village government) in performing certain development functions. Following the cooperative reforms of the liberalisation era, the PCSs became partly detached from state political and economic structures, and in some cases this resulted in 'disempowerment' of the PCSs, or at least some loss of the power to effect certain political outcomes at the village level. The recently acquired ability to gain access to resources from the tobacco-processing companies is likely to 're-empower' the PCS leadership as one of the key agencies mediating community development vis-à-vis village government. This obviously affects the range and content of socio-economic issues that the PCSs are supposed to address, and the abilities of the PCS leadership to handle development affairs for local communities will necessarily have a deep and often decisive influence on the locally specific impacts or results of a development project. In some cases, for example, SRPs projects are being used 'pragmatically' by PCSs to serve their own interests, such as financial support to acquire new furniture and electronic devices for the PCS office and PCS board members. In other cases, the realisation of development projects has caused tensions at the village level. For example, one of the PCSs (PCS5, Table 13.2) received 3 million Tsh from one of the tobacco companies in 2009 to rehabilitate the public primary school in the area. Almost immediately after the smooth completion of the rehabilitation, severe conflicts provoked by massive side-selling practices by more than half the PCS tobacco farmers resulted in their expulsion from the PCS. In 2014, the school had not been operating for several years, apparently because the remaining members of the PCS refused to put their children 'under the same roof' as peers from households that had violated the contract.

While socio-economic differentiation between tobacco farmers within PCSs is nothing new in the tobacco-growing area, processes of differentiation are likely to be reinforced by the tobacco-processing companies' agenda for CSR compliance (for an elaboration of emerging social differentiation during the *Ujamaa*, pre- and post-liberalisation periods, see Boesen and Mohele, 1979, and Ponte, 2002). The farmer packages are intentionally provided only to well-performing farmers and present new opportunities for improving the livelihoods of some farmers within the PCSs, often the wealthier ones. One likely consequence might be that income distribution within the PCS will become more concentrated in favour of the richer tobacco-growing households, who are able not only to improve their housing condition and accumulate more household assets, including land and cattle, but also to invest in off-farm activities as part of a livelihood diversification strategy. However, the longer-term influence on local patterns of socio-economic differentiation and rural capitalist development still remains to be seen.

Concluding remarks

At the beginning of the new millennium, the major international cigarette branders became increasingly exposed to allegations of negative labour practices and the massive deforestation of woodlands associated with tobacco-growing areas in developing countries. In order to secure brand reputation, the branded cigarette manufacturers adapted and 'greened' their value chain practices by integrating social and environmental concerns into their industry-wide CSR programmes. In this context, social and environmental concerns have become important elements of governance in the global tobacco value chain, while an increasing number of performance requirements are being imposed by cigarette branders on their leaf suppliers in tobacco-growing regions. This chapter has demonstrated the recent embeddedness of social and environmental concerns in tobacco contract farming schemes in Tanzania. The tightly coordinated contract farming schemes and the strategic alliance between the two tobacco-processing companies to coordinate and monitor smallholders and PCSs ensures crucial on-farm compliance with good agricultural practices (including deforestation) and the prevention of child labour especially. This is consolidated through the contractual agreements between the tobacco-processing companies and the PCSs. The requirements to demonstrate good agricultural practices and labour practices are thoroughly addressed in the contents of the contract, while ATTT staff rigorously monitor and train farmers in CSR requirements.

The tobacco processing companies are also increasingly making voluntary financial contributions to local community development projects, co-financed by the branded cigarette manufacturers as part of their CSR programmes. Through SRPs and farmer packages, resources are allocated to loyal and well-performing PCSs and, within PCS communities, to individual farmers. The local development projects accompany the tobacco companies' strategic focus on achieving a reliable supply base and at the same time addressing and/or rewarding the compliance of PCSs with crucial CSR requirements. Most of the SRPs have direct development impacts on local communities through in-kind contributions to the construction and/or maintenance of public goods such as school facilities or improved access to clean water, while the farmer packages contribute to furthering the well-being of selected farmers and are directly linked to the production process. These types of community assistance with the aim of fostering local socio-economic development mark a partial shift from public responsibilities for local development to private-sector-driven development initiatives in the main tobacco-growing regions in Tanzania. This raises questions regarding the impacts on local communities and individuals and the decisive influence on local patterns of socio-economic development, including emerging social and spatial inequalities, all of which requires further research.

Notes

1 The chapter is based on fieldwork conducted in 2013 and 2014 in the main tobacco-producing region in Tanzania. Interviews were conducted with the two tobacco-processing companies (headquartered in Morogoro), and their regional (Tabora) and district (Urambo) representatives. In addition, interviews were conducted with the Association of Tanzanian Tobacco Traders (representatives of the regional headquarters in Tabora and district headquarters in Urambo). In 2013, interviews were conducted with ten primary cooperative societies and in 2014 with 11 primary cooperative societies. These interviews were supplemented by focus-group interviews with farmers held within the same primary cooperative society.

2 For a concise and interesting overview of the 'battlefield' during the dark ages of the tobacco industry, see Palazzo and Richter (2005).

3 The International Union of Food, Agriculture, Hotel, Restaurant, Catering, Tobacco and Allied Workers' Associations.

4 It is notable that leading international opponents such as the WHO continue to question the very possibility of 'social responsibility' in the tobacco industry and categorise the CSR programmes as an 'inherent contradiction', as the industry's main business activity runs counter to the 'societal good' (Palazzo and Richter, 2005).

5 The other processing plant, located in Songea, was sold to a Cooperative Union and a foreign investor, but this factory has not been modernised (Mitchell and Baffes, 2002).

6 Although the input credit system has been beset with problems since the contractual arrangements were revamped, it released the tobacco companies from the financial burden of tying up working capital in the input distribution system (see Chapter 4, this volume, on the recent restructuring of the contractual system in tobacco).
7 According to Otañez and Glantz (2011), production of 1 kg of tobacco requires 20 kg of firewood for curing. This is equivalent to 10–40 tons of dry wood per ton of processed tobacco (see also Mangora, 2012).

References

Alliance One (2016). One Vison of Action-oriented Social Responsibility. Online: www.aoisustainability.com. Accessed 25 May 2016.

Alliance One International (2016). Online: www.aointl.com. Accessed 25 May 2016.

Boesen, J. and Mohele, A. T. (1979). *The 'Success Story' of Peasant Tobacco Production in Tanzania: The Political Economy of Commodity Producing Peasantry*. Uppsala: Scandinavian Institute of African Studies.

ECLT (Eliminating Child Labour in Tobacco Growing) (2013a). Child Labour in Tobacco Growing. Online: www.eclt.org/good-practices-for-eliminating-child-labour-in-agriculture. Accessed 25 May 2016.

ECLT (2013b). Promoting Sustainable Practices to Eradicate Child Labour in Tobacco. Online: www.eclt.org/project-countries/tanzania/tanzania-2004-2006. Accessed 25 May 2016.

ECLT (2014). Pledge of Commitment by ECLT Foundation Board Members. Online: www.eclt.org. Accessed 25 May 2016.

ECLT (2015). Members' Pledge of Commitment. Online: www.eclt.org/about-us/members-pledge-commitment. Accessed 25 May 2016.

Geist, H., Chang, K. T., Etges, V. and Abdallah, J. (2009). Tobacco Growers at the Crossroads: Towards a Comparison of Diversification and Ecosystem Impacts. *Land Use Policy*, 26(4), 1066–1079.

Gonzalez, M., Green, L. W. and Glantz, S. A. (2011). Through Tobacco Industry Eyes: Civil Society and the FCTC Process from Philip Morris and British American Tobacco's Perspectives. *Tobacco Control*, 21(4).

Jamil, D. (2010). The CSR of MNC Subsidiaries in Developing Countries: Global, Local, Substantive or Diluted. *Journal of Business Ethics*, 93, 181–200.

Jamil, D. and Naville, B. (2011). Convergence versus Divergence of CSR in Developing Countries: An Embedded Multi-Layered Institutional Lens. *Journal of Business Ethics*, 102, 599–621.

Japan Tobacco International (2016). Online: www.jti.com/brands. Accessed 25 May 2016.

Mangora, M. M. (2012). Shifting Cultivation, Wood Use and Deforestation Attributes of Tobacco Farming in Urambo District, Tanzania. *Current Research Journal of Social Science*, 4(2), 135–140.

Mitchell, D. and Baffes, J. (2002). *Tanzanian Agricultural Export: Challenges and Constraints in a Global Economy*. Washington, DC: World Bank.

Otañez, M. and Glantz, S. A. (2011). Social Responsibility in Tobacco Production? Tobacco Companies' Use of Green Supply Chains to Obscure the Real Costs of Tobacco Farming. *Tobacco Control*, 20(6), 403–411.

Otañez, M., Muggli, M. E., Hurt, R. D. and Glantz, S. A. (2006). Eliminating Child Labour in Malawi: A British American Tobacco Corporate Responsibility Project to Sidestep Tobacco Labour Exploitation. *Tobacco Control*, 15, 224–230.

Palazzo, G. and Richter, U. (2005). CSR Business as Usual? The Case of the Tobacco Industry. *Journal of Business Ethics*, 61(4), 387–401.

Philip Morris (2016). Responsibility at Philips Morris. Online: www.altria.com/our-companies/philipmorrisusa/Pages/default.aspx. Accessed 25 May 2016.

Ponte, S. (2002). *Farmers and Markets in Tanzania*. Oxford: James Currey.

Tobacco Buying Agreement (n.d.). *Tobacco Buying Agreement between Primary Cooperative Society Limited (Seller) and Cooperative Union and Buyer*. Obtained from Alliance One, Morogoro, July 2013.

Universal Leaf Tobacco (2016). Online: www.universalcorp.com/UniversalLeaf. Accessed 25 May 2016.

Index

Page numbers in *italics* denote tables, those in **bold** denote figures.

For Product Safety Concerns and Information please contact our EU representative GPSR@taylorandfrancis.com Taylor & Francis Verlag GmbH, Kaufingerstraße 24, 80331 München, Germany